AF602152

Plant Analytical Techniques

The Authors

Dr S. D. Ramteke, Principal Scientist, ICAR-National Research Centre for Grapes, Pune. He has more than 25 years experience in plant physiology. He did B.Sc. (Forestry), M.Sc. (Agril. Botany) from PKV, Akola, received Ph.D (Crop Physiology) from University of Agril. Sciences, Dharwad. He has published more than 90 research papers, 200 popular articles, covering 5 book chapters/reviews, 6 books. He is recipient of many award. Dr Ramteke visited South Africa, U.S. and China for paper presentation.

Dr Jayant Haridas Meshram, presently working as Senior Scientist (Plant Physiology) with at ICAR- Central Institute for Cotton Research, Nagpur. He did Ph.D in Plant Physiology from IARI, New Delhi. He has published 20 research papers in international journals of repute, edited books and several popular articles. He visited USA and Brazil research conference.

Plant Analytical Techniques

Compiled by

Dr S. D. Ramteke
Department of Plant Physiology
ICAR-National Research Centre for Grapes,
Pune-412307, Maharashtra

&

Dr Jayant Haridas Meshram

Assisted by

Mahadev Bhoje
Ram Kar
Ujjwala Jape
Pankaj Bankar

2019

Daya Publishing House®
A Division of
Astral International Pvt. Ltd.
New Delhi – 110 002

ISBN: 9789388173971 (Int. Edition)

Publisher's Note:

Every possible effort has been made to ensure that the information contained in this book is accurate at the time of going to press, and the publisher and author cannot accept responsibility for any errors or omissions, however caused. No responsibility for loss or damage occasioned to any person acting, or refraining from action, as a result of the material in this publication can be accepted by the editor, the publisher or the author. The Publisher is not associated with any product or vendor mentioned in the book. The contents of this work are intended to further general scientific research, understanding and discussion only. Readers should consult with a specialist where appropriate.

Every effort has been made to trace the owners of copyright material used in this book, if any. The author and the publisher will be grateful for any omission brought to their notice for acknowledgement in the future editions of the book.

Published by : **Daya Publishing House®**
A Division of
Astral International Pvt. Ltd.
– ISO 9001:2015 Certified Company –
4736/23, Ansari Road, Darya Ganj
New Delhi-110 002
Ph. 011-43549197, 23278134
E-mail: info@astralint.com
Website: www.astralint.com

Digitally Printed at : **Replika Press Pvt. Ltd.**

Preface

Most of the plant scientists experience difficulty in getting proper analytical methods for their research needs and they refer various books and journals in library. Methods are available but the time required to search for techniques needed for their research is tremendous. The research workers sometimes give up the essential measurements for want of methods and such data they feel would have added more clarity in their findings. Every researcher goes through such problems. At that time there are few textbook available that could the serve the needs of researchers to find the best methods in a single book. This book provides abundant analytical procedures a research worker needs in analyzing the plant samples as per objectives of targeted projects during short preparations.

The time of a worker is precious because he has to complete the project during stipulated period. Most of the procedures needed by plant physiologists, phytochemists, taxonomists, environmentalists, food technologists, toxicologists, tissue analysts, horticulturists and soil chemists are specifically included in this book which will definitely help researchers in programming their projects more efficiently and satisfactorily. The procedures for tissue analysis of nutrient deficiencies in plant samples are dire needs of the field workers. This shall certainly facilitate quick diagnosis for accurate recommendations to the farmers.

It is known that the precise information cannot be obtained without the analytical methods. The developments of analytical techniques in the past decade have necessitated compiling all the precise methods in a single book. This book serves the purpose of thousands of researchers and field workers who can obtain the required procedures for basic and advanced field studies as well. I am sure that the students shall surely benefit in project planning for synopsis because some critical analyses are of prime importance to find missing links to enhance the scientific concept. This book is perhaps the first of its kind where large numbers of methods are available in one cover with preparatory instructions at primary level, which I feel is the utmost need of the day.

Finally we wish to again emphasize that this book will help both guides and students as well to unearth the hidden information for successful interpretation and proper discussion. I hope the book meets the expectations of all the plant scientists and shall certainly have better patronage in India and abroad.

Dr S. D. Ramteke

Contents

1

Chapter

Extraction and Quantifications Procedures of Carbohydrates

Introduction

Carbohydrates are principal compounds in a plant body. Chemically these are hydrates of carbon and have the general formula as C (H_2O) n. They exist in different forms, which are congenitally classified into three major classes, viz., Monosaccharide's, Oligosaccharides and Polysaccharides. The simplest soluble forms of carbohydrates are 3- carbon compounds, glyceraldehydes, and dihydroxyacetone. All carbohydrates are polyhydroxy - aldehydes or ketones. Polysaccharides constitute the most complex form of carbohydrates of high molecular weight. Monosaccharide contains the free aldehyde or ketone group. Some disaccharides have the free aldehyde group (maltose) and some do not have the free ones (sucrose). The polysaccharides, starch, and cellulose are the polymers of a monosaccharide linked through the active groups. Among the carbohydrates Sucrose is one of the main photosynthetic products in higher plants and is the main form in which carbohydrates are translocated from the leaves to the rest of the plan parts, supplying these organs with carbon and energy essential for growth. In addition, this sugar is the main carbohydrate stored in some plants, appearing in leaves, branches, tubers, and fruits. On the other hand, sucrose is highly important in plants for acclimation to environmental stress caused by cold, salt and drought.

In view of the above, it is necessary to perform a quantification study on the influence that different levels of carbohydrates (from deficiency to toxicity) exert on the metabolism and distribution of non-structural carbohydrates in all of the plants.

The aim of the caring out such an analysis is to determine whether these quantification studies can be considered good indicators of carbohydrates sufficiency.

1.1 Color Reaction of Carbohydrates

The chemical properties of saccharides vary depending upon the number of hydroxyl groups and the presence or absence of –CHO/= CO groups. These variations are the basis in the development of color reactions to identify the saccharides. Some simple tests used to identify the presence/absence of certain saccharides are listed below:

Reagents

- *a)* ***Iodine solution*** Add a few crystals of iodine to 2% potassium iodide solution till the color becomes deep yellow.
- *b)* ***Fehling's reagent A:*** Dissolve 34.65 g copper sulphate in distilled water and make up to 500 ml.
- *c)* ***Fehling's reagent B:*** Dissolve 125 g potassium hydroxide and 173 g Rochelle salt (potassium sodium tartrate) in distilled water and make up to 500 ml.
- *d)* ***Benedict's qualitative reagent:*** Dissolve 173 g sodium citrate and 100 g sodium carbonate in about 500 ml water. Heat to dissolve the salts and filter, if necessary. Dissolve 17.3 g copper sulphate in about 100 ml water and add it to the above solution with stirring and make up the volume to 1 L with water.
- *e)* ***Barfoed's reagent:*** Dissolve 24 g copper acetate in 450 ml boiling water. Immediately add 25 ml of 8.5% lactic acid to the hot solution. Mix well, Cool and dilute to 500 ml.
- *f)* ***Seliwanoff's reagent:*** Dissolve 0.05 g resorcinol in 100 ml dilutes (1:2) hydrochloric acid.
- *g)* ***Bial's reagent:*** Dissolve 1.5 g orcinol in 500 ml of concentrated HCl and add 20 to 30drops of 10% ferric chloride.

The reactions of carbohydrates are given in Table 1.1.

Experiment	Observation	Remarks
Molisch's Test Add two drops of Molisch's reagent (5% 1-naphthol in alcohol) to about 2 ml of test solution and mix well. Incline the tube and add about1 ml of concentrated sulphuric acid along the sides of the tube. Observe the color at the junction of the two liquids.	A red-cum-violet ring appears at the junction of the two liquids.	The color formed is due to the reaction of alpha-naphthol with furfural and/or its derivatives formed by the dehydration of sugars by concentrated sulphuric acid. All carbohydrates react positively with this reagent.
2. ***Iodine Test*** Add a few drops of iodine solution to about 1 ml of the test solution.	Appearance of deep blue color.	This indicates the presence of starch in the solution. The blue color is due to the form action of starch - iodine complex.

3. *Fehling's Test* To 1 ml of Fehling's solution 'A', add 1 ml of Fehling's solution 'B' and a few drops of the test solution. Boil for a few minutes.	Formation of yellow or brownish-red precipitate.	The blue alkaline cupric hydroxide present in Fehling's solution, when heated in the presence of reducing sugars, gets reduced to yellow or red cuprous oxide and it gets precipitated. Hence, formation of the colored precipitate indicates the presence of reducing sugars in the test solution.
4. *Benedict's Test* To 2 ml of Benedict's reagent add five drops of the test solution. Boil for five minutes in a water bath. Cool the solution.	Formation of red, yellow or green color/ precipitate.	As in Fehling's test, the reducing sugars because of having potentially free aldehyde or keto group reduce cupric hydroxide in alkaline solution to red colored cuprous oxide. Depending on the sugar concentration yellow to green color is developed.

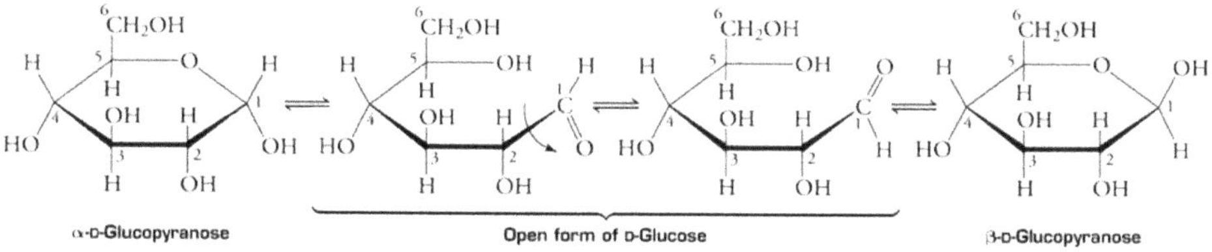

α-D-Glucopyranose ⇌ Open form of D-Glucose ⇌ β-D-Glucopyranose

5. *Barfoed's Test:* To 1 ml of the test solution add about 2 ml of Barfoed's reagent. Boil it for one minute and allow to 6. Stand for a few minutes.	Formation of brick-red precipitate.	Only monosaccharide's answer this test. Since Barfoed's reagent is weakly acidic, it is reduced only by monosaccharide's.
6. *Seliwanoff's Test* To 2 ml of Seliwanoff's reagent add two drops of test solution and heat the mixture to just boiling.	Appearance of deep red color.	In concentrated HCl, ketoses undergo dehydration to yield furfural derivatives more rapidly than do aldoses. These derivatives form complexes with resorcinol to yield deep red color. It is a timed color reaction specific for ketoses.

7. ***Bial's Test*** To 5 ml of Bial's reagent add 2–3 ml of solution and warm gently. When bubbles rise to the surface cool under the tap.	Appearance of green color or precipitate.	It is specific for pentoses. They get converted to furfural. In the presence of ferric ion orcinol and furfural condense to yield a colored product.
8. ***Test for non-reducing sugars such as sucrose:*** (*a*) Do Benedict's test with the test solution. (*b*) Add 5 drops of concentrated HCl to 5 ml of test solution in another test tube. Heat for five minutes on a boiling water bath. Add 10% sodium hydroxide solution to give a slightly alkaline solution (test with red litmus paper). Now perform Benedict's test with this hydrolyzed solution.	No characteristic color formation. Appearance of red or yellow color.	Indicates the absence of reducing sugars in the given solution. Indicates the formation of reducing sugars from non-reducing sugars after hydrolysis with acid.
9. ***Mucic Acid Test*** Add a few drops of conc. HNO_3 to the concentrated test solution or substance directly and evaporate it over a boiling water bath till the acid fumes are expelled. Add a few drops of water and leave it overnight.	Formation of crystals.	The both end carbon groups are oxidized to carboxylic groups. The resultant saccharic acid of galactose is called mucic acid which is insoluble in water.
10. ***Osazone Test*** To 0.5 g of phenylhydrazine - hydrochloride add 0.1 g of sodium acetate and 10 drops of glacial acetic acid. To this mixture add 5 ml of test solution and heat on a boiling water bath for about half an hour. Allow the tube to cool slowly and examine the crystals under a microscope.	Glucose, fructose and mannose produce needle-shaped yellow osazone crystals, whereas lactosazone is mushroom shaped. Different osazones show crystals of different shapes. Maltose produces flower-shaped crystals.	The ketoses and aldoses react With phenylhydrazine to produce a phenylhydrazone which in turn reacts with another two molecules of phenylhydrazine to form the osazone.

Notes

1. For ozone test, the reaction mixture should be between pH 5 and 6. Fructose takes 2 min to form the osazone whereas for glucose it is 5 min.

The disaccharides take a longer time to form osazones. Disaccharides form crystals only on cooling.

2. When a mixture of carbohydrates is present in the test sample, chromatographic methods should be employed to identify the individual sugars.

1.2 Determination of Reducing Sugars by Nelson-Somogyi Method

Sugars with reducing property (arising out of the presence of a potential aldehyde or keto group) are called reducing sugars. Some of the reducing sugars are glucose, galactose, lactose, and maltose. The Nelson-Somogyi method is one of the classical and widely used methods for the quantitative determination of reducing sugars.

Principle

The reducing sugars when heated with alkaline copper tartrate reduce the copper from the cupric to cuprous state and thus cuprous oxide is formed. When the cuprous oxide is treated with arsenomolybdic acid, the reduction of molybdic acid to molybdenum blue takes place. The blue color developed is compared with a set of standards in a colorimeter at 620 nm.

Materials

- Alkaline Copper Tartrate:
  - i) Dissolve 2.5 g anhydrous sodium carbonate, 2 g sodium bicarbonate, 2.5 g potassium sodium tartrate and 20 g anhydrous sodium sulphate in 80 ml water and makeup to 100 ml.
  - ii) Dissolve 15 g copper sulphate in a small volume of distilled water. Add one drop of sulphuric acid and makeup to 100 ml. Mix 4 ml of B and 96 ml of solution A before use.
- Arsenomolybdate reagent:

Dissolve 2.5 g ammonium molybdate in 45 ml water. Add 2.5 ml sulphuric acid and mix well. Then add 0.3 g disodium hydrogen arsenate dissolved in 25 ml water. Mix well and incubate at 37°C for 24–48 hours.

- Standard glucose solution:

  Stock: 100 mg in 100 ml distilled water.
- Working standard:

  10 ml of stock diluted to 100 ml with distilled water [100μg/ml].

Procedure

1. Weigh 100 mg of the sample and extract the sugars with hot 80% ethanol twice (5 ml each time).
2. Collect the supernatant and evaporate it by keeping it in a water bath at 80°C.
3. Add 10 ml water and dissolve the sugars.
4. Pipette out aliquots of 0.1 or 0.2 ml to separate test tubes.

5. Pipette out 0.2, 0.4, 0.6, 0.8 and 1 ml of the working standard solution into a series of test tubes.
6. Make up the volume to both sample and standard tubes to 2 ml with distilled water.
7. Pipette out 2 ml distilled water in a separate tube to set a blank.
8. Add 1 ml of alkaline copper tartrate reagent to each tube.
9. Place the tubes in boiling water for 10 minutes.
10. Cool the tubes and add 1 ml of arsenomolybolic acid reagent to all the tubes.
11. Make up the volume in each tube to 10 ml with water.
12. Read the absorbance of blue color at 620 nm after 10 min.
13. From the graph drawn, calculate the amount of reducing sugars present in the sample.

Calculation

Absorbance corresponds to 0.1 ml of test = x mg of glucose

10 ml contains = $\frac{X}{0.1} \times 10$ mg of glucose

= % of reducing sugars

1.3 Estimation of Reducing Sugar by Dinitrosalicylic Acid Method

For sugar estimation an alternative to Nelson-Somogyi method is the dinitrosalicylic acid method. It's simple, sensitive and adoptable during handling of a large number of samples at a time.

Materials

- **Dinitrosalicylic Acid Reagent (DNS Reagent)**

  Dissolve by stirring 1 g dinitrosalicylic acid, 200 mg crystalline phenol and 50 mg sodium sulphite in 100 ml 1% NaOH. Store at 4°C. Since the reagent deteriorates due to sodium sulphite, if long storage is required, sodium sulphite may be added at the time of use.
- 40% Rochelle salt solution (Potassium sodium tartrate).

Procedure

1. Follow, steps 1 to 3 as in Nelson-Somogyi's method to extract the reducing sugars from the test material.
2. Pipette out 0.5 to 3 ml of the extract in test tubes and equalize the volume to 3 ml with water in all the tubes.
3. Add 3 ml of DNS reagent.
4. Heat the contents in a boiling water bath for 5 min.
5. When the contents of the tubes are still warm, add 1 ml of 40% Rochelle salt solution.

6. Cool and read the intensity of dark red color at 510 nm.
7. Run a series of standards using glucose (0–500 μg) and plot a graph.

Calculation

Calculate the amount of reducing sugars present in the sample using the standard graph.

1.4 Determination of Non Reducing Sugars

Principle

Non-reducing sugars are characterized by the absence of an open chain structure, so they're not susceptible to oxidation-reduction reactions. The most common non-reducing sugar is sucrose. Benedict's test is important for testing the presence of non reducing sugar in food or solution. Benedict's reagent consists of anhydrous sodium carbonate, sodium citrate and copper (II) sulfate pentahydrate. Collectively, this blue-colored reagent serves as an oxidizing agent that will change colors if a reaction occurs. Non-reducing sugars do not react with the solution in an oxidation-reduction reaction because they do not have an open chain structure.

Reagents

Benedict's Qualitative Reagent: Dissolve 173 sodium citrate and 100g sodium carbonate in about 800 mL water. Heat to dissolve the salts and filter, if necessary.

Dissolve 17.3g copper sulphate in about 100mL water and add it to the above solution with stirring and make up the volume to 1L with water.

Procedure

1. To 2ml of Benedict's reagent add five drops of the test solution.
2. Boil for five minutes in a water bath.
3. Cool the solution. This makes the solution appear as a blue color.
4. For the presence of non-reducing sugars, the solution should remain the same blue color for approximately 10 minutes. If the solution begins to change to a red color, it indicates the presence of reducing sugars.

1.5 Determination of Glucose by Glucose Oxidize Method

Glucose is a widely distributed simple sugar with an active aldehyde group. Estimation of glucose by glucose oxidase gives the true glucose concentration eliminating the interference by other reducing sugars.

Principle

Glucose oxidase catalyses the oxidation of alpha-D -glucose to D-glucono 1, 5 lactone (gluconic acid) with the formation of hydrogen peroxides. The oxygen liberated from hydrogen peroxide by peroxidase reacts with the O-dianisidine and oxidises it to a red chromophore product.

$$\text{Glucose} + O_2 \xrightarrow[\text{Oxidation}]{\text{Glucose}} H_2O_2 + \text{Gluconic Acid}$$

$$H_2O_2 + \text{O-dianisidine} \xrightarrow{\text{Peroxidase}} \text{Red-coloured product}$$

Materials

- **Glucose Oxidase Peroxidase Reagent**

  Dissolve 25 mg O-dianisidine completely in 1 ml of methanol. Add 49 ml of 0.1 Mphosphate buffer (pH 6.5). Then add 5 mg of peroxidase and 5 mg of glucose oxidase tothe above prepared O-dianisidine solution.

- **Standard Dissolve 100 mg glucose in 100 ml water.**

  Dilute 10 ml of this stock to100 ml to obtain the working standard.

Procedure

1. Pipette out o.5 ml of deproteinised plant extract (deproteinization is not necessary in samples with very low protein content)
2. Add 0.5 ml distilled water and 1 ml glucose oxidase peroxidase reagent.
3. Into a series of test tubes pipette out 0 (blank), 0.2, 0.4, 0.6, 0.8 and 1 ml of working standard glucose solution and make up the volume to 1.0 ml with distilled water. Then add 1 ml of glucose oxidase -peroxidase reagent.
4. Incubate all the tubes at 35°C for 40 minutes.
5. Terminate the reaction by the addition of 2 ml of 6 N-HCl.
6. Read the color intensity at 540 nm.

Calculation

From the standard graph, calculate the amount of glucose present in the sample.

1.6 Determination of Total Carbohydrates by Anthrone Method

Carbohydrates are the important components of storage and structural materials in the plants. They exist as free sugars and polysaccharides. The basic units of carbohydrates are the monosaccharide's which cannot be split by hydrolysis into simpler sugars. The carbohydrate content can be measured by hydrolyzing the polysaccharides into simple sugars by acid hydrolysis and estimating the resultant mono saccharides.

Principle

Carbohydrates are first hydrolyzed into simple sugars using dilute hydrochloric acid. In hot acidic medium glucose is dehydrated to hydroxyl methyl furfural. This compound forms with anthrone green colored product with an absorption maximum at 630 nm.

Materials

- **2.5 N HCl**
- **Anthrone reagent:**

  Dissolve 200 mg anthrone in 100 ml of ice-cold 95% H_2SO_4. Prepare fresh before use.

- **Standard glucose:**

  Stock solution: -Dissolve 100 mg in 100 ml of distilled water.

Working standard—10 ml of stock diluted to 100 ml with distilled water. Store refrigerated after adding a few drops of toluene.

Procedure

1. Weigh 100 mg of the sample into a boiling tube.
2. Hydrolyze by keeping it in a boiling water bath for three hours with 5 ml of 2.5 N HCl and cool to room temperature.
3. Neutralize it with solid sodium carbonate until the effervescence ceases.
4. Make up the volume to 100 ml and centrifuge.
5. Collect the supernatant and take 0.5 and 1 ml aliquots for analysis.
6. Prepare the standards by taking 0, 0.2, 0.4, 0.6, 0.8 and 1 ml of the working standard. '0' serves as blank.
7. Make up the volume to 1 ml in all the tubes including the sample tubes by adding distilled water.
8. Then add 4 ml of anthrone reagent.
9. Heat for eight minutes in a boiling water bath.
10. Cool rapidly and read the green to dark green color at 630 nm.
11. Draw a standard graph by plotting concentration of the standard on the X-axis *versus* absorbance on the Y-axis.
12. From the graph calculate the amount of carbohydrate present in the sample tube.

Calculation

Amount of carbohydrate present in 100 mg of the sample

$$= \frac{\text{Mg of glucose}}{\text{Volume of test sample}} \times 100$$

Note

Cool the contents of all the tubes on ice before adding ice-cold anthrone reagent.

1.7 Phenol Sulphuric Acid Method for Total Carbohydrates

The phenol sulphuric acid method to estimate total carbohydrates is described below.

Principle

In hot acidic medium glucose is dehydrated to hydroxyl methyl furfural. This forms a green colored product with phenol and has absorption maximum at 490 nm.

Reaction Theory

In the presence of strong acids and heat, carbohydrates undergo a series of reactions that leads to the formation of furan derivatives such as furan-aldehyde and hydroxyl methyl furaldehyde. The initial reaction, a dehydration reaction is followed by the formation of furan derivatives, which then condense with themselves or Phenolic compounds to produce dark colored complexes. Displays some furan derivatives and the carbohydrates from which they originate. The developed complex absorbs UV-VI light, and the absorbance is proportional to the sugar concentration in a linear fashion. An absorbance maximum is observed at 490nm for hexoses and 480nm for pentose's and urosnic acids as measured by a UV-VI spectrophotometer

Materials

- **Phenol 5%:**

  Redistilled (reagent grade) phenol (50 g) dissolved in water and diluted to one liter
- **Sulphuric acid 96% reagent** grade
- **Standard glucose**

  Stock—100 mg in 100 ml of water,
- **Working standard**

  10 ml of stock diluted to 100 ml with distilled water.

Procedure

1. Follow the steps 1 to 4 as given in anthrone method for sample preparation.
2. Pipette out 0.2, 0.4, 0.6, 0.8 and 1 ml of the working standard into a series of test tubes.
3. Pipette out 0.1 and 0.2 ml of the sample solution in two separate test tubes. Make up the volume in each tube to 1 ml with water.
4. Set a blank with 1 ml of water.
5. Add 1 ml of phenol solution to each tube.
6. Add 5 ml of 96% sulphuric acid to each tube and shake well.
7. After 10 min shake the contents in the tubes and place in a water bath at 25–30°C for 20 min.
8. Read the color at 490 nm.
9. Calculate the amount of total carbohydrate present in the sample solution using the standard graph.

Calculation

Absorbance corresponds to 0.1 ml of the test

$$= x \text{ mg of glucose}$$

100 ml of the sample solution contains

$$= \frac{X}{0.1} \times 100 \text{ mg of glucose}$$

= % of total carbohydrate present.

1.8 Estimation of Starch by Anthrone Reagent

Starch is an important polysaccharide. It is the storage form of carbohydrate in plants abundantly found in roots, tubers, stems, fruits and cereals. Starch, which is composed of several glucose molecules, is a mixture of two types of components namely amylose and amylopectin. Starch is hydrolyzed into simple sugars by dilute acids and the quantity of simple sugars is measured calorimetrically.

Principle

The sample is treated with 80% alcohol to remove sugars and then starch is extracted with perchloric acid. In hot acidic medium starch is hydrolyzed to glucose and dehydrated to hydroxyl methyl furfural. This compound forms a green colored product with anthrone.

Materials

- **Anthrone**

  Dissolve 200 mg anthrone in 100 ml of ice-cold 95% sulphuric acid.
- **80% ethanol.**
- **52% perchloric acid.**
- **Standard glucose**:

  Stock—100 mg in 100 ml water
- Working standard:

10 ml of stock diluted to 100 ml with water

Procedure

1. Homogenize 0.1–0.5 g of the sample in hot 80% ethanol to remove sugars. Centrifuge and retain the residue. Wash the residue repeatedly with hot 80% ethanol till the washings do not give color with anthrone reagent. Dry the residue well over a water bath.
2. To the residue add 5.0 ml of water and 6.5 ml of 52% perchloric acid.
3. Extract at 0°C for 20 min. Centrifuge and save the supernatant.
4. Repeat the extraction using fresh perchloric acid. Centrifuge and pool the supernatants and make up to 100 ml.
5. Pipette out 0.1 or 0.2 ml of the supernatant and make up the volume to 1 ml with water.
6. Prepare the standards by taking 0.2, 0.4, 0.6, 0.8 and 1 ml of the working standard and make up the volume to 1 ml in each tube with water.
7. Add 4 ml of anthrone reagent to each tube.
8. Heat for eight minutes in a boiling water bath.
9. Cool rapidly and read the intensity of green to dark green color at 630 nm.

Calculation

Find out the glucose content in the sample using the standard graph. Multiply the value by a factor 0.9 to arrive at the starch content.

1.9 Determination of Amylose

Starch is composed of two components, namely amylose and amylopectin. Amylose is a linear or non-branched polymer of glucose. The glucose units are joined by α-1-4 glucosidic linkages. Amylose exists in coiled form and each coil contains six glucose residues.

Principle

The iodine is adsorbed within the helical coils of amylose to produce a blue-colored complex which is measured colorimetrically.

Materials

- Distilled water
- Ethanol.
- 1 N NaOH.
- 0.1% phenolphthalein.
- Iodine reagent:

  Dissolve 1 g iodine and 10 g KI in water and make up to 500 ml.
- Standard:

  Dissolve 100 mg amylose in 10 ml 1 N NaOH; make up to 100 ml with water.

Procedure

1. Weigh 100 mg of the powdered sample, and add 1 ml of distilled ethanol. Then add10 ml of 1 N NaOH and leave it overnight.
2. Make up the volume to 100 ml.
3. Take 2.5 ml of the extract, add about 20 ml distilled water and then three drops of phenolphthalein.
4. Add 0.1 N HCl drop by drop until the pink color just disappears.
5. Add 1 ml of iodine reagent and make up the volume to 50 ml and read the color at 590 nm.
6. Take 0.2, 0.4, 0.6, 0.8 and 1 ml of the standard amylose solution and develop the color as in the case of sample.
7. Calculate the amount of amylose present in the sample using the standard graph.
8. Dilute 1 ml of iodine reagent to 50 ml with distilled water for a blank.

Calculation

Absorbance corresponds to 2.5 ml of the test solution

$$= x \text{ mg amylose}$$

100 ml contains

$$= \frac{X}{2.5} \times 100 \text{ mg amylose.}$$

= % amylose

Notes

1. The sample suspension may be heated for 10 min in a boiling water-bath instead of overnight dissolution.
2. The amount of amylo-pectin is obtained by subtracting the amylose content from that of starch.

References

1.10 Estimation of Cellulose

Cellulose, a major structural polysaccharide in plants, is the most abundant organic compound in nature, and is composed of glucose units joined together in the form of the repeating unit's of the disaccharide cellobiose with numerous cross linkages. It is also a major component in many of the farm wastes.

Principle

Cellulose undergoes acetolysis with acetic/nitric reagent forming acetylated cellodextrins which get dissolved and hydrolyzed to form glucose molecules on treatment with 67% H_2SO_4. This glucose molecule is dehydrated to form hydroxymethyl furfural which forms green colored product with anthrone and the color intensity is measured at 630 nm.

Materials

- **Acetic/Nitric reagent**

  Mix 150 ml of 80% acetic acid and 15 ml of concentrated nitric acid.
- **Anthrone reagent**

  Dissolve 200 mg anthrone in 100 ml concentrated sulphuric acid. Prepare fresh and chill for 2 h before use.
- 67% Sulphuric acid.

Procedure

1. Add 3 ml acetic/nitric reagent to a known amount (0.5 g or 1 g) of the sample in a test tube and mix in a vortex mixer.
2. Place the tube in a water-bath at 100°C for 30 min.
3. Cool and then centrifuge the contents for 15–20 min.
4. Discard the supernatant.
5. Wash the residue with distilled water.
6. Add 10 ml of 67% sulphuric acid and allow it to stand for 1 h.
7. Dilute 1 ml of the above solution to 100 ml.
8. To 1 ml of this diluted solution, add 10 ml of anthrone reagent and mix well.
9. Heat the tubes in a boiling water-bath for 10 min.
10. Cool and measure the color at 630 nm.
11. Set a blank with anthrone reagent and distilled water.

12. Take 100 mg cellulose in a test tube and proceed from Step No. 6 for standard.

(Instead of just taking 1 ml of the diluted solution (Step 7) take a series of volumes (say 0.4–2 ml corresponding to 40–200 μg of cellulose) and develop the color).

Calculation

Draw the standard graph and calculate the amount of cellulose in the sample.

1.11 Estimation of Hemi Cellulose

Hemicelluloses are non-cellulosic, non-pectic cell wall polysaccharides. They are regarded as being composed of xylans, mannans, glucomannans, galactans and arabinogalactans. Hemicelluloses are categorized under 'unavailable carbohydrates' since they are not split by the digestive enzymes of the human system.

Principle

Refluxing the sample material with neutral detergents solution removes the water-soluble and materials other than the fibrous components. The left out material is weighed after filtration and expressed as Neutral Detergent Fiber (NDF).

Materials

- Neutral Detergent Solution

  Weigh 18.61 g disodium ethylenediaminetetraacetate and 6.81 g sodium borate- decahydrate. Transfer to a beaker. Dissolve in about 200 ml of distilled water by heating and to this, add a solution (about 100–200 ml) containing 30 g of sodium lauryl sulphate and 10 ml of 2-ethoxy ethanol. To this add a solution (about 100ml) containing 4.5g of disodium hydrogen phosphate. Make up the volume to one liter and adjust the pH to 7.0.

- Decahydronaphthalene.
- Sodium sulphite.
- Acetone

Procedure

1. To 1 g of the powdered sample in a refluxing flask add 10 ml of cold neutral detergent solution.
2. Add 2 ml of decahydronaphthalene and 0.5 g sodium sulphite.
3. Heat to boiling and reflux for 60 min.
4. Filter the contents through sintered glass crucible (G-2) by suction and wash with hot water.
5. Finally give two washings with acetone.
6. Transfer the residue to a crucible, dry at 100°C for 8 h.
7. Cool the crucible indesiccators and weigh.

Calculation

Hemi cellulose= Neutral detergent fiber (NDF) – Acid detergent fiber (ADF)

Note

See Lignin for determining acid detergent fiber.

1.12 Determination of Fructose and Inulin

Fructose, a ketohexose (called as fruit sugar), is usually accompanied by sucrose in fruits like apple. Honey is a rich source of fructose.

1.12.1 Determination of Fructose

Principle

The hydroxyl methyl furfural formed from fructose in acid medium reacts with resorcinol to give a red color product.

Reaction Theory

Fructose can also be quantitatively assayed using hexokinase. Glucose and fructose can be assayed together using hexokinase, glucose-6-phosphatedehydrogenase, and phosphoglucoseisomerase (PGI) to catalyze specific reactions. In the presence of ATP and hexokinase, glucose and fructose are phosphorylated to glucose-6-phosphate and fructose-6-phosphate (F-6-P) respectively. Adding NAD and glucose-6-phosphate dehydrogenase oxidizes G-6-P to gluconate-6-phosphate and results in the formation of NADH which can be measured at 340 nm. Adding phospho-glucose isomerase changes F-6-P into G-6-P, which is then oxidized to gluconate-6-phosphate in the presence of NAD and leads to the formation of NADH. Sucrose may also be assayed using hexokinase or glucose oxidase by first treating it with invertase to release glucose and fructose.

Materials

- Resorcinol reagent

  Dissolve 1 g resorcinol and 0.25 g thiourea in 100 ml glacial aceticacid. This solution is indefinitely stable in the dark.

- Dilute HCl

  Mix five parts of conc. HCl with one part of distilled water.

- Standard fructose solution

  Dissolve 50 mg of fructose in 50 ml water.

  Dilute 5 ml of this stock to 50 ml for a working standard.

Procedure

1. To 2 ml of the solution containing 20–80 μg of fructose adds 1 ml of resorcinol reagent.
2. Then add 7 ml of dilute hydrochloric acid.
3. Pipette out 0.2, 0.4, 0.6, 0.8 and 1 ml of the working standard and make up the volume to 2 ml with water. Add 1 ml of resorcinol reagent and 7 ml of dilute HCl as above.

4. Set a blank along with the working standard.
5. Heat all the tubes in a water-bath at 80°C for exactly 10 min.
6. Remove and cool the tubes by immersing in tap water for 5 min.
7. Read the color at 520 nm within 30 min.
8. Draw the standard graph and calculate the amount of fructose present in the sample using the standard graph.

1.12.2 Determination of Inulin

Inulin is a polymer made of fructose units with β-2-1 linkage. It is found in onion, garlic and in many other plant parts.

Sample Extraction

Grind the sample and extract in 80% ethanol for six hours to remove free sugars. Dry the sample and take 500 mg in a 100 ml conical flask. Add 20 ml of water and heat it in a water bath at 90°C for 10 min. Collect the extract and then add 70 ml of water. Replace the flask for another 30 min with occasional shaking to dissolve the fructosan, then remove and cool it at room temperature. Combine the extracts and filter the solution if it is not clear and make up to100 ml in a standard flask. To estimate the inulin content in the extract follow the procedure given for fructose estimation. The amount of inulin is expressed in terms of fructose concentration.

1.13 Estimation of Pectin Substances

Pectic substances abundantly exist in the middle lamella of the plant cells. There are three types of pectic substances—pectic acids, pectin and proto-pectin. Pectic acid is un branched molecule made up of about 100 units of D-galacturonic acid residues. The monomers are linked through 1–4 linkages. Pectin is an extensively esterifies pectic acid. Several carboxyl groups exist as methyl esters. Pectic acid is water soluble whereas pectin forms a colloidal solution. Proto-pectin is a larger molecule than pectic acid and pectin. During ripening of fruits, conversion of proto-pectin into pectic acid and pectin takes place. The pectin's in fruits vary in their methoxyl content and in jellying power.

Two methods are described below for the estimation of pectin: one gravimetric and the other, colorimetric.

1.13.1 Gravimetric Method

Principle

Pectin is extracted from plant material and saponified. It is precipitated as calcium pectate by the addition of calcium chloride to an acid solution. After thoroughly washing to eliminate chloride ions, the precipitate is dried and weighed.

Materials

- **1 N Acetic acid :**

  Dilute 30 ml of glacial acetic acid to 500 ml with water.

- **1 N Calcium chloride solution**

  Dissolve 27.5 g anhydrous $CaCl_2$ in water and dilute to 500 ml.

- 1% Silver nitrate:

  Dissolve 1 g $AgNO_3$ in 100 ml water.
- 0.01 N HCl
- 0.05 N HCl
- 0.3 N HCl

Procedure

1. Weigh 50 g of blended sample into a 1 L beaker and add 300 ml 0.01 N HCl. Boil for 30 min and filter under suction. Wash the residue with hot water and collect the filtrate.
2. To the residue add 100 ml 0.05 N HCl, boil for 20 min filter, wash and collect the filtrate.
3. To the residue now add 100 ml 0.3 N HCl; boil for 10 min, filter, wash and collect the filtrate.
4. Pool the filtrates. Cool and make to volume (500 ml).
5. Pipette out 100–200 ml aliquots into 1 L beakers.
6. Add 250 ml water and neutralize the acid with 1 N NaOH using phenolphthalein indicator. Add an excess of 10 ml of 1 N NaOH with constant stirring and allow it to stand overnight.
7. Add 50 ml 1 N acetic acid and after 5 min, add 25 ml 1 N calcium chloride solution with stirring. Allow it to stand for 1 h.
8. Boil for 1 to 2 min.
9. Filter through a pre-weighed Whatman No. 1 filter paper (see note 1).
10. Wash the precipitate with almost boiling water until the filtrate is free from chloride.
11. Test the filtrate with silver nitrate for chloride.
12. Transfer the filter paper with the calcium pectate, dry overnight at 100°C in a weighing dish, cool in a desiccators and weigh.

Calculation

The pectin content is expressed as % calcium pectate

$$\%\ \text{calcium pectate} = \frac{\text{Wt. of calcium pectate}}{\text{ml of filtrate taken} \times \text{Wt. of sample for estimation}}$$

Notes

The filter paper for Step No. 9 should be prepared as described below:

1. Wet the filter paper in hot water, dry in oven at 102°C for 2 h. Cool in desiccators and weigh in a covered dish. The theoretical yield of calcium pectate from pure galacturonic anhydride is 110.6%.

1.13.2 Colorimetric Method

Principle

Galacturonic acid is reacted with carbazole in the presence of H_2SO_4 and the color developed is measured at 520 nm.

Materials

- 60% Ethyl alcohol :

  Mix 500 ml 95% alcohol and 300 ml water.
- 95% Ethyl alcohol.
- Purified ethyl alcohol:

  (Reflux 1 L of 95% ethyl alcohol with 4 g zinc dust and 2 ml conc. H_2SO_4 for 15 h and distill in all glass distillation apparatus. Redistill with 4 g zinc dust and 4 g KOH).
- 1 N and 0.05 N Sodium hydroxide
- H_2SO_4 (Analytical grade)
- 0.1% Carbazole reagent:

  Weigh 100 mg re-crystallized Carbazole, dissolve and dilute to 100 ml with purified alcohol.

Procedure

1. Weigh 100 mg pectin (see notes section for the preparation of pectin) and dissolve in 100 ml of 0.05 N NaOH.
2. Allow it to stand for 30 min to de-esterify the pectin.
3. Take 2 ml of this solution and make up to 100 ml with water.
4. Pipette out 2 ml of de-esterifies pectin solution and add 1 ml carbazole reagent. A white precipitate will be formed.
5. Add 12 ml conc. H_2SO_4 with constant stirring.
6. Close the tubes with rubber stopper and allow standing for 10 min to develop the color.
7. To set a blank add 1 ml of purified ethyl alcohol in the place of carbazole reagent.
8. Read the color at 525 nm against blank, exactly 15 min after the addition of acid.

Standard

Weigh 120.5 mg Galacturonic acid monohydrate (from a sample vacuum dried for 5 h at 30°C) and transfer to a 1 liter volumetric flask. Add 10 ml 0.05 N NaOH and dilute to volume with water. After mixing, allow it to stand overnight. Dilute 10, 20, 40, 50, 60 and 80 ml of this standard solution to 100 ml with water. Take 2ml of these solutions for color developing and proceed as in the case of the sample. Draw a standard curve with the absorbance *versus* concentration.

Calculation

Read the concentration of the anhydrogalacturonic acid corresponding to the reading of the sample, and calculate as follows:

$$\% \; anhydrogalacturonic\ acid = \frac{\mu g\ of\ anhydrogalacturonic\ acid\ in\ the\ aliquot \times Dilution \times 100}{ml\ taken\ for\ estimation \times Wt.\ of\ pectin\ sample \times 1{,}000{,}000}$$

Notes

1. Carbazole is re-crystallized from toluene.
2. An alternate procedure adopted for color development is as follows:

Take 12 ml of conc. H_2SO_4 in a test tube, cools in an ice-bath, and add 2 ml of de-esterifies pectin solution and again cool. Heat the contents in a boiling water-bath for 10 min, cool to 20°C and add 1 ml of 0.15% carbazole reagent in purified ethyl alcohol. Allow it to stand for 25 ± 5 minute at room temperature to develop the color. Read the absorbance at 520 nm. Standards should also be treated similarly.

1.14 Extraction and Purification of Pectin

1. Blend the fresh sample. If the material is dry grind
2. Transfer 100 g macerated sample (10 g dry tissue) to a pre-weighed 1 L beaker containing 400 ml water.
3. Add 1.2 g freshly ground sodium hexametaphosphate and adjusts to pH 4.5.
4. Heat with stirring at 90–95°C for 1 h. Check the pH in every 15 min and maintain at pH5 with citric acid or NaOH. Replace water lost by evaporation at intervals. However,do not add water at the last 20 min.
5. Add 4 g filter aid and 4 g ground paper pulp. Filter rapidly through a fast filter papercoated with 3 g moistened fast filter aid.
6. Collect at least 200 ml of the filtrate in a pre-weighed container. Cool as rapidly as possible. Now, note the weight of the filtrate.
7. If the filtrate contains less than 0.2% pectin, concentrate the filtrate under vacuum to attain this concentration.
8. To three volumes of ethanol, iso -propanol or acetone containing 0.5 N HCl, pour the cooled, weighed filtrate. The slurry should be at pH 0.7–1. Stir it for 30 min.
9. Centrifuge or filter. Wash the precipitate with the same solvent containing HCl. Then, wash repeatedly with 70% alcohol or acetone until the precipitate is essentially chloride free or the pH is above 4.
10. Dehydrate the precipitate further in 400 ml acetone. Dry it for an overnight *in a vacuum* with a slow stream of dry air passing through the oven.
11. Weigh the precipitate and use this pectin for analysis.

12. The dried pectin should be free from ammonia for which a small sample of the pectin is heated with 1 ml of 0.1 N NaOH and ammoniacal odor can be noticed or tested with a moistened litmus paper. If ammonium ions are present wash with acidified 6% alcohol, followed by neutral alcohol to remove the acid and dry.

1.15 Estimation of Crude Fiber

Crude fiber consists largely of cellulose and lignin (97%) plus some mineral matter. It represents only 60–80% of the cellulose and 4–6% of the lignin. The crude fiber content is commonly used as a measure of the nutritive value of poultry and livestock feeds and also in the analysis of various foods and food products to detect adulteration, quality and quantity.

Principle

During the acid and subsequent alkali treatment, oxidative hydrolytic degradation of the native cellulose and considerable degradation of lignin occur. The residue obtained after final filtration is weighed, incinerated, cooled and weighed again. The loss in weight gives the crude fiber content.

Materials

- Sulphuric acid solution (0.255 ± 0.005 N):

  1.25 g concentrated sulphuric acid diluted to 100 ml (concentration must be checked by titration).
- Sodium hydroxide solution (0.313 ± 0.005 N): 1.25 g sodium hydroxide in 100 ml distilled water (concentration must be checked by titration with standard acid).

Procedure

1. Extract 2 g of ground material with ether or petroleum ether to remove fat (Initial boiling temperature 35–38°C and final temperature 52°C). If fat content is below 1%, extraction may be omitted.
2. After extraction with ether boil 2 g of dried material with 200 ml of sulphuric acid for 30 min with bumping chips.
3. Filter through muslin and wash with boiling water until washings are no longer acidic.
4. Boil with 200 ml of sodium hydroxide solution for 30 min.
5. Filter through muslin cloth again and wash with 25 ml of boiling 1.25% H_2SO_4, three 50 ml portions of water and 25 ml alcohol. Remove the residue and transfer to ashing dish (pre-weighed dish W1).
6. Dry the residue for 2 h at 130 ± 2°C. Cool the dish in desiccators and weigh (W2).
7. Ignite for 30 min at 600 ± 15°C.
8. Cool in desiccators and reweigh (W3).

Calculation

$$\% \text{ crude fiber in ground sample} = \frac{\text{Loss in weight on ignition } (W_2 - W_1) - (W_3 - W_1)}{\text{Weight of the sample}} \times 100$$

1.16 Estimation of Pyruvic Acid

Pyruvic acid or pyruvate is an important metabolic intermediate. It is greatly produced in the terminal step of glycolysis and funnels to TCA cycle for further oxidation for releasing the chemical energy. It can be determined following the procedure gıven below:

Principle

The DNPH (2, 4-dinitrophenyl hydrazine) reacts with pyruvate after the addition NaOH giving a brown colored hydrazone product which can be estimated calorimetrically at 510 nm.

Materials

- **Phosphate buffer pH** 9.4

  A: 0.2 M solution of monobasic sodium phosphate $NaH_2PO_4H_2O$ (27.8 g in 1000 ml).

  B: 0.2 M solution of dibasic sodium phosphate (53.65 g of $Na_2HPO_4.7H_2O$ in 1 L or 17.7g of $Na_2HPO_4.12H_2O$ in 1 L). 19 ml of A and 81 ml of B, diluted to a total of 200 ml Store in refrigerator.

- **Pyruvate, Standard**

  Dissolve 22 mg sodium pyruvate in 100 ml water in a standard flask.

- 2, **4-Dinitrophenyl hydrazin**e (DNPH)

  Dissolve 19.8 mg of DNPH in 10 ml of conc. HCl and make to 100 ml with water.

  Store it in an amber bottle at room temperature

- Sodium hydroxide 0.8 N

  Dissolve 16 g sodium hydroxide in one liter water.

- **Plant extract**

  Grind 6 g of plant material in 15 ml of phosphate buffer. Centrifuge at 25,000 g for 15min. Use the supernatant as plant extract.

Procedure

1. Pipette out 50 μl, 75 μl, 100 μl, 150 μl, 200 μl of pyruvate standard solution and 0.5, 1.0 ml, 1.5 ml, and 2.0 ml of sample extract into test tubes and make up the volume to 2.0 ml with phosphate buffer (pH 7.4).
2. Set a blank with no pyruvate solution.
3. Add 0.5 ml of DNPH solution to each tube.
4. Incubate at 37°C for 20–30 min.

5. Add 5 ml of NaOH solution to each tube mix well and incubate for 10 min at room temperature.
6. Record the absorbance at 610 nm.
7. Draw the standard graph and calculate the amount of pyruvic acid present in the sample using the graph.

1.17 Determination of Fruit Acids by Titration

It is the sugar/acid ratio which contributes towards giving many fruits their characteristic flavor and so is an indicator of commercial and organoleptic ripeness. At the beginning of the ripening process the sugar/acid ratio is low, because of low sugar content and high fruit acid content, this makes the fruit taste sour. During the ripening process the fruit acids are degraded, the sugar content increases and the sugar/acid ratio achieves a higher value. Over ripe fruits have very low levels of fruit acid and therefore lack characteristic flavor.

Titration is a chemical process used in ascertaining the amount of constituent substance in a sample, e.g. acids, by using a standard counter-active reagent, e.g. an alkali (NaOH).

Once the acid level in a sample has been determined it can be used to find the ratio of sugar to acid.

There are two methods specified for the determination of the titratable acidity of fruits:

a) Method using a colored indicator;
b) Potentiometric method, using a pH meter, which should be used for very colored juices.

Material

- A laboratory burette of 25 or 50 ml capacity or an automatic burette is used. A 10ml pipette, beaker (250 ml), a filter (muslin cloth or fine filter) and an extractor or homogenizer.
- A bottle of distilled water
- **Sodium Hydroxide (NaOH):**

  The Standard Laboratory solution of 0.1M which is used in the actual titration is considered to be dilute, and can readily be purchased in this form.
- Phenolphthalein:

  This is a 1% w/v solution of phenolphthalein in 95% v/v ethanol which is flammable and toxic if ingested. This is only required for the method using a colored indicator.
- Indicator stripes:

To check the exact point of neutrality an indicator stripe should be used. Not necessary if pH Meter is used.

Sampling

To evaluate the lot selected for inspection, take a sample of at least 10 fruits of each size at random from the reduced sample. However, fruits should be free from defects such as sun scorch and pest or disease damage, which may have affected the normal ripening process.

Sample Preparation

Depending upon the type of produce, either cut the fruit in half and squeeze out the juice with an extractor or a juice-press e.g. citrus fruits, or homogenize the flesh into a pulp. The juice of all squeezed fruits is mixed.

The skin and solids should not be included; the solids being filtered out through muslin cloth or fine filter extracting as much juice as possible. Use a clean and dry safety 10ml pipette. Draw up 10 ml of juice and discharge it into a 250 ml beaker. Using another clean and dry pipette draw up 50 ml of distilled water and add to the juice in the beaker.

Measurement

Method Using a Colored Indicator

Add 3 drops of phenolphthalein to the juice/water solution in each beaker from a dropping pipette which is specifically kept for that purpose.

Ensure the tap on the burette is shut and using a funnel pour the 0.1M solution of NaOH into the burette until it reaches the zero mark. Do not spill the solution onto the skin.

Slowly titrate with the NaOH into the juice/water solution (with a 25ml burette or an automatic burette). Care must be taken that the NaOH is dropped directly into the solution and does not adhere to the glass; otherwise the reading may be false. At the same time as titrating care must be taken to continually swirl the solution in the beaker to keep the solution thoroughly mixed. This is essential, particularly when the solution nears neutrality. It is important to determine the point of neutrality or the end point of titration very exactly. The phenolphthalein indicator changes very rapidly from colorless to pink and the end point can easily be missed, which will give an inaccurate reading for the test. It is important therefore that towards the end of the titration the NaOH is added a drop at a time.

Using phenolphthalein as an indicator, the point of neutrality is reached when the indicator changes from colorless to pink. The indicator color must remain stable (persisting for 30 seconds) and be light pink when viewed over a white background. However, the shade can vary depending on the type of juice being tested. If the point of neutrality is missed, i.e. the color of the indicator is too dark; the test is not acceptable and must be repeated. An indicator stripe should be used to avoid the neutral point of pH 8.1.

- Read the amount of the amount of NaOH used (titer) on the burette and record this figure.
- Re-fill the burette for each subsequent test.
- Clean the equipment thoroughly and rinse with distilled water. Detergents must not be used.

Note

When testing very acidic juices *e.g.* lemons and limes a larger amount of NaOH is required. Therefore, when the NaOH reaches the 25 ml mark on the scale the burette tube should be recharged as described above. When the end point is reached the various readings are added together and recorded to produce a figure of NaOH used for each titration.

Method Using a pH Meter

The point of neutrality i.e. the end point of titration may also be determined using a pH meter. The precise method used will depend on the manufacturer instructions, but the following will provide a general guide.

Checking the pH Meter

- Make sure the pH meter has warmed up before use - allow about 30 minutes.
- Remove the electrode from the distilled water in the storage beaker and dry.
- Place the electrode into the beaker containing a buffer solution of pH 7 and calibrate the meter to the same figure.
- Whenever readings are taken, ensure that the electrode is not in contact with the sides or base of the beaker.
- Remove the electrode and - after rinsing in distilled water - place in the solution to be tested; the electrode should not have any contact with the glass.

Measurement

1. Ensure the tap on the burette is shut and using a funnel pour the 0.1M solution of NaOH into the burette until it reaches the zero mark. Do not spill the solution onto the skin.
2. Slowly titrate with the NaOH into the juice/water solution. Care must be taken that the NaOH is dropped directly into the solution and does not adhere to the glass; otherwise the reading may be false. While titrating care must be taken to continually stir the solution in the beaker to keep thoroughly mixed. This is essential, particularly when the solution nears neutrality. It is important to determine the point of neutrality or the end point of titration very exactly. The end point can easily be missed, which will give an inaccurate reading for the test. It is important therefore that towards the end of the titration the NaOH is added a drop at a time.
3. Using a pH meter, while titrating the digital readout will be seen to climb from around 4 or 5. When the reading reaches 7 proceed carefully. The point of neutrality or the end point of titration is reached at pH 8.1. If this figure is exceeded the test is not acceptable and must be repeated.
4. When the pH meter reads 8.1 read off the amount of NaOH used on the burette and record.

5. Remove the electrode and rinse it in distilled water ready for the next test. Do not allow it to become contaminated.
6. Re-fill the burette for each subsequent test.
7. Clean the equipment thoroughly and rinse with distilled water. Detergents must not be used.

Note

When testing very acidic juices e.g. lemons and limes a larger amount of NaOH is required. Therefore, when the NaOH reaches the 25 ml mark on the scale the burette tube should be recharged as described above. When the end point is reached the various readings are added together and recorded to produce a figure of NaOH used for each titration.

1.18 Calculation of the Sugar/Acid Ratio

The °Brix value of the fruit concerned must also be obtained before calculation of the sugar/acid ratio is possible.

The calculations for determining the sugar/acid ratios of all produce are the same, but as some products contain different acids the appropriate multiplication factor must be applied to each calculation. Some products may contain more than one type of acid; it is the primary acid that is tested. A list of these acids and multiplication factors are found below.

Factor for:

- Citric acid: 0.0064 (Citrus fruit)
- Malic acid: 0.0067 (Apples)
- Tartaric acid: 0.0075 (Grapes)

Using citric acid as an example, 1ml 0.1M NaOH is equivalent to 0.0064g citric acid.

Results

Results expressed as percentage acid:

$$\text{Percentage acid} = \frac{\text{Titer x acid factor} \times 100}{10\ (\text{ml juice})}$$

$$\text{The sugar acid ratio} = \frac{^\circ\text{Brix value}}{\text{Percentage acid}}$$

OR

Results

Results expressed as acid in gram/liter:

$$\text{g/l acid} = \frac{\text{Titer x acid factor} \times 100 \times 10}{10\ (\text{ml juice})}$$

$$\text{The sugar acid ratio} = \frac{^\circ\text{Brix value} \times 10}{\text{g/l acid}}$$

E.g. In case of citric acid the result would be expressed as

Percentage citric acid	gram/ liter acid
$\text{Percentage Citric Acid} = \frac{\text{Titer} \times 0.0064 \times 100}{10 \text{ ml juice}}$	$\text{g/l Citric Acid} = \frac{\text{Titer} \times 0.0064 \times 100 \times 10}{10\,(\text{ml juice})}$
This formula can be simplified to: Percentage Citric Acid = Titer x 0, 064	This formula can be simplified to: g/l Citric Acid = Titer x 0, 64
$\text{The sugar acid ratio} = \frac{°\text{Brix value}}{\text{Percentage acid}}$	$\text{The sugar Acid Ratio} = \frac{°\text{Brix value} \times 10}{\text{g/l Citric Acid}}$

Results

- It is important to record the results, to one decimal place, as well as all the details concerning method, variety and stage of maturity and ripeness of the produce being tested.
- If the result achieves the limit specified in the standard, the lot has reached the minimum maturity level.
- If the result is at least 10 per cent below/above the limit specified in the standard, a second sample needs to be taken and analyzed with other fruits of the reduced sample or from a new sample.
- If the average of the two samples is below/above the limit specified in the standard, the lot fails the minimum maturity level and needs to be rejected. No tolerance is applied.

Health and Safety Guidelines During Titration

Sodium hydroxide in its undiluted form is extremely corrosive to body tissue. Skin contact causes irritation almost immediately and continued contact causes burns. The 0.1 Molar solution used in this test is much safer. However, it is recommended that protective coats are worn when using, and that it is used only in a well ventilated room.

Phenolphthaleinis highly flammable and should be used with care. It should be stored and used away from naked flames or other sources of ignition. It is toxic if ingested.

References

Ashwell G (1957) In: Methods in Enzymol 3 (Eds Colowick, S J and Kaplan N O) Academic Press New York p 75.

Dubois M., Gilles K A., Hamilon J K., Rebers P A and Smith F (1956) *Anal Chem* 26 350.

Goering H D and Vansoest P J. (1975) Forage fiber analysis US Dept of Agriculture Agricultural Research Service Washington.

Hedge J E and Hofreiter B T (1962) In: Carbohydrate Chemistry17 (Eds Whistler R L and Be Miller J N) Academic Press New York.

Hedge J E and Hofreiter B T (1962) In: Carbohydrate Chemistry 17 (Eds *Whistler R* L and Be Miller J N) Academic Press New York.

http://brilliantbiologystudent.weebly.com/benedicts-test-for-non-reducing-sugars.html

http://citeseerx.ist.psu.edu/viewdoc/download?doi=10.1.1.334.9409&rep=rep1&type=pdf

http://insa.nic.in/writereaddata/UpLoadedFiles/PINSA/Vol12_1946_1_Art03.pdf

http://insa.nic.in/writereaddata/UpLoadedFiles/PINSA/Vol12_1946_1_Art03.pdf

http://journals.sagepub.com/doi/pdf/10.1177/000456326900600108

http://onlinelibrary.wiley.com/doi/10.1002/0471142913.fae0203s00/abstract?userIsAuthenticated=false&deniedAccessCustomisedMessage=

http://onlinelibrary.wiley.com/doi/10.1002/jsfa.2740070108/full

http://onlinelibrary.wiley.com/doi/10.1002/star.19960480103/full

http://onlinelibrary.wiley.com/doi/10.1111/j.1365-2621.1953.tb17701.x/full

http://pubs.acs.org/doi/abs/10.1021/jf60116a018?journalCode=jafcau

http://technologyinscience.blogspot.in/2011/04/determination-of-reducing-sugars-by.html#.Wg0vN4-CzIU

http://technologyinscience.blogspot.in/2011/04/estimation-of-reducing-sugar-by.html#.Wg0wb4-CzIU

http://www.biochemden.com/qualitative-analysis-of-carbohydrates/

http://www.biocyclopedia.com/index/plant_protocols/carbohydrates/color_reactions_of_carbohydrates.php

http://www.biocyclopedia.com/index/plant_protocols/carbohydrates/reducing_sugars_by_Nelson-Somogyi_method.php

http://www.flavours.asia/uploads/7/9/8/9/7989988/oecd_fruit_acid_determination_method.pdf

http://www.labsynergy.com/titration-of-acidity-in-fruit-juices/

http://www.scielo.br/scielo.php?script=sci_arttext&pid=S0100-29452015000400835

http://www.tandfonline.com/doi/abs/10.1080/15567030902937127

http://www.tandfonline.com/doi/abs/10.1080/15567030902937127

https://doubledotthree.wordpress.com/2014/10/15/benedict-test-for-reducing-and-non-reducing-sugar-biology/

https://link.springer.com/article/10.1007/s00217-011-1552-3

https://link.springer.com/article/10.1007/s11274-006-9258-6

https://link.springer.com/chapter/10.1007/978-1-4419-1463-7_6

https://link.springer.com/chapter/10.1007/978-94-011-3700-3_4

https://www.cabdirect.org/cabdirect/abstract/19481400014

https://www.ncbi.nlm.nih.gov/pmc/articles/PMC1259056/

https://www.ncbi.nlm.nih.gov/pmc/articles/PMC1264364/

https://www.ncbi.nlm.nih.gov/pmc/articles/PMC1264364/

https://www.ncbi.nlm.nih.gov/pmc/articles/PMC1269789/

Juliano B O (1971) CerealSci Today 16 334

Krishnaveni, S, Theymoli Balasubramanian and Sadasivam S (1984) Food Chem15 229.

Krishnaveni S, Theymoli Balasubramanian and Sadasivam S (1984) Food Chem15 229.

Krishnaveni S., Theymoli Balasubramanian and Sadasivam S. (1984) Food Chem 15 229.

Malik C P and Singh M B (1980) Plant enzymology and Histoenzymology Kalyani Publisher New Delhi p 278.

Maynard A J (Ed) (1970) Methods in Food Analysis Press New York p 176.

McCready R M, Guggolz J, Siliviera V and Owens H S (1950) Anal Chem 22 1156.

Miller G L (1972) Anal Chem31 426.

Somogyi M (1952) J biol Chem200 245.

Thayumanavan B and Sadasivam S (1984) Plant foods Hum Nutr 34 253.

Thayumanavan B and Sadasivam S (1984) Plant foods Hum Nutr 34 253.

Updegroff D M (1969) Anal Biochem 32 420.

2

Chapter

Extraction and Quantifications Procedures of Proteins and Amino Acids

2.1 Assays for Total Proteins

Introduction

Proteins are large biological molecules consisting of one or more chains of amino acids. Proteins perform a vast array of functions within living organisms, including catalyzing metabolic reactions,replicating DNA,responding to stimuli, and transporting molecules from one location to another. Proteins differ from one another primarily in their sequence of amino acids, which is dictated by the nucleotide sequence of their genes, and which usually results in folding of the protein into a specific three-dimensional structure that determines its activity.

A polypeptide is a single linear polymer chain of amino acids bonded together by peptide bonds between the carboxyland amino groups of adjacent amino acid residues. The sequence of amino acids in a protein is defined by the sequence of agene, which is encoded in the genetic code. Shortly after or even during synthesis, the residues in a protein are often chemically modified by post translational modification, which alters the physical and chemical properties, folding, stability, activity, and ultimately, the function of the proteins. Sometimes proteins have non-peptide groups attached, which can be called prosthetic groups or cofactors. Proteins can also work together to achieve a particular function, and they often associate to form stable protein complexes. Therefore the protein determination is necessary to estimate the amount of protein in the sample, to normalize against the protein concentration or during purification procedures. Depending on the amount of sample, accuracy and presence of interfering agents, one needs to decide on the method to be used. For accurate quantification, the sample protein is compared with a known amount of a standard protein which could either be the commonly used bovine serum albumin (BSA) or it could sometimes be immunoglobulin G (IgG). The various methods and their specifications are outlined below:

2.1.1 Biuret Assay

Principle

The biuret reaction is based on the complex formation of cupric ions with proteins. In this reaction, copper sulphate is added to a protein solution in strong alkaline solution. A purplish-violet color is produced, resulting from complex formation between the cupric ions and the peptide bond. The name of the reaction is derived from biuret, which forms a similar colored complex with cupric ions. The biuret reaction with proteins is independent of the composition of the protein; therefore protein composition is not a factor. However, protein purity and association state could influence the results obtained with the biuret reagent.

In general, biuret assays are useful for samples containing -1 to 10 mg protein / ml, which is diluted -5-fold by the added reagent to give a concentration of 0.2 to 2 mg/ml final assay volume (f.a.v.). Most proteins produce a deep purple color, with maximum absorbance (L,)at 550 nm.

Materials

- **Reagents**

  Weigh 1.50 gm of cupric sulfate ($CuSO_4$. 5 H_2O) and 6.0 gm. Of sodium potassium tartrate ($NaKC_4H_2O$. $4H_2O$); transfer to a dry 1 liter volumetric flask, and dissolve in about 500 ml of water. Add with constant swirling 300 ml. of 10 percent sodium hydroxide solution). Make to volume with water, mix and store in a paraffin-lined bottle. This reagent should keep indefinitely but must be discarded if, as a result of contamination or faulty preparation. It shows sign of depositing any black or reddish precipitation.

- **Calibration standard**

  For *e. g.* 10 to 20 mg/ ml BSA. Buffer or solvent used to prepare the protein-containing sample.

- Sample containing protein at 1 to 10 mg/d. Biuret reagent
- Spectrophotometer and I-cm cuvettes

Procedure

1. Prepare a dilution series of calibration standard in the buffer or solvent used to prepare the sample.
   a) Use bovine serum albumin (crystallized or lyophilized or one of the Cohn Fraction V. Preparations which are 96% to 98% protein and 3% to 4% water; Sigma) for the calibration standard Concentrations of albumin in the final assay volume (f.a.v.) of 5 ml may range from -0.40 to 2.50 mg protein/ml-to produce A_{550}readings from -0.10 to 0.70 in I-cm. cuvettes-so the dilution series should run from 2 to 12.5 mg/ml.
2. In separate test tubes, add 1 ml of protein-containing sample, of each dilution of the calibration standard, or of the buffer or solvent used to prepare the sample (reference standard) to 4ml biuret reagent. Incubate 20 min at room temperature.
   a) The modest amounts of detergent (*e.g.*, deoxycholate or SDS) used to help solubilize proteins from tissue or insoluble biomaterials that will be added with the protein sample have negligible effects on this assay, although they sometimes have marked effects on the Hanree-Lowry

assay Mild reducing agents and strong oxidizers adversely affect the assay].

3. Measure the net absorbance of the sample, calibration standards, and reference standard at 550 nm (A) in 1-ml cuvettes. If the spectrophotometer does not automatically give net absorbance readings, subtract the value of the reference standard from those obtained for the sample and calibration standards.
   a) [Sample should not be diluted after the addition of biuret reagent. In general, reaction mixtures for spectrophotometric assays should not be diluted after development even if the color (absorbance) is too intense. Dilution will decrease the required excess of cupric ion, thus upsetting chemical euilibria. If the sample has an A_{550}>1 or 2(depending on the spectrophotometer), dilute a sample of the original protein and repeat the assay].
4. Prepare a calibration plot by graphing the net A_{550}values for the standards versus protein concentration (μg protein/ml f. a. v.). Determine the protein concentration of the sample by interpolation from the plot.
   a) [Plot the data and calculate the slope and its units. (The plots in this unit show net absorbance as the ordinate.) The slope of this plot is directly proportional to the apparent spectrophotometric absorption coefficient (the sensitivity of the assay) in the same units].

Results

The calibrations curves we have obtained using the reagent and the method describe above obey satisfactorily the law of Lambert and Beer.

2.1.2 Hartree-Lowry Assay

The original Lowry method for total protein analysis was first described in one of the most cited papers in biochemistry (Lowry *et al.*, 1951). The assay is a colorimetric assay based on cupric ions and Foli -Ciocalteau reagent for Phenolic groups. The assay has been re-investigated many times and sometimes improved. Most of these studies were designed to discern how interfering compounds distort the assay and how detergents solubilize otherwise insoluble proteins. The literature related to this assay was comprehensively reviewed by Peterson (1983). This protocol describes Hartree's version of the Lowry assay (Hartree, 1972). This version uses three reagents instead of five, produces more intense color (increased sensitivity) with some proteins, maintains a linear response over a larger concentration range (30% to 40% greater), is less easy to overload, and overcomes the reagent salt co -precipitation encounter with Lowry reagents. Finally, the reagents formulation offers some advantages in storage stability. It is somewhat less laborious than the original Lowry assay, and it maintains the sensitivity of the original.

Principle

The Lowry reaction is based on the amplification of the biuret reaction by subsequent reaction with the Folin phenol reagent (Folin–Ciocalteu reagent).

Factors other than the biuret reaction play a role in the development of color in the Lowry method, resulting in considerable variation with respect to protein composition. This variation is a reflection of the contributions of specific amino acids (tyrosine, tryptophan) on color development in this reaction. The Lowry assay has been used to determine the protein level 25 to 150 μg of protein in the sample at 660 nm.

Materials

- Calibration standard: 300 pg crystalline BSA per ml in water
- Buffer or solvent used to prepare the protein-containing sample
- Sample containing protein at 100 to 600 μg/ml
- Hartree-Lowry reagents A, B, and C
- 50°C water bath
- Spectrophotometer and
- 1-cm cuvettes

Procedure

1. Prepare a dilution series of calibration standard in the same buffer or solvent used to prepare the sample to give concentrations of 30 to 150 pg /ml.
   a) Concentrations of albumin in the final assay volume (F. a .v.) of 5 ml may range from-30 to 300 pg protein/5 m1(F. a. v.) to produce A_{650} readings from -0.20 to 0.80.
2. Add 1.0 ml of the protein-containing sample, of each dilution of calibration standard, or of the buffer or solvent used to prepare the sample (reference standard) to 0.90 ml of Hartree-Lowry reagent in separate test tubes. Incubate 10 min in a 50°C water bath.
3. Cool the tubes to room temperature.
4. Add 0.1 ml of Hartree-Lowry reagent B to each tube and mix. Incubate 10 min at room temperature.
5. Rapidly add 3ml Hartree-Lowry reagent C to each tube and mix thoroughly. Incubate 10 min in a 50°C water bath, and then cool to room temperature.
   a) The final assay volume is 5.0 ml.
6. Measure the net absorbance the sample, calibration standards, and reference standard at 650 nm (A) in 1-cm cuvettes. If the spectrophotometer does not automatically give net absorbance readings, subtract the value for the reference solution from those obtained for the sample and calibration standards.
7. Prepare a calibration plot by graphing the net A650 values for the standards versus protein concentration (pg protein15 ml f. a. v.).Determine the protein concentration of the sample by interpolation from the plot.

a) Plot the data as shown in Figure 3.4.2 and calculate the slope and its units. The slope of a calibration plot is proportional to the spectrophotometric absorption coefficient, and is a measure of the sensitivity of assay over the range indicated on the horizontal axis. Comparisons can be made with other measures, such as the weight absorption coefficient ($E^{1\%}$) at the same wavelength. Units in the concentration factor in thedenominator of ($E^{1\%}$) are (dry g)/100 ml of final assay volume.

8. The units in this protocol, μg protein/ml final assay volume (f. a.v.), are used to permit comparison with values in the literature. It is not always clear whether the milligram or microgram amounts of protein indicated on the areas of these plots refer to the concentration in the input sample or to the final concentration after the addition of reagent(s). Differences involve rimes of input versus final developed samples frequently involve dilution factors of 10′ to 102. As an example, the $E^{1\%}$ values in the original report describing this procedure Lowry *et al.*, (1951) are -230 at 750 nm with 3-mm path lengths. These values can be converted to units presented in this protocol by reconciling the two means for expressing the slopes.

$$E^{1\%}_{1cm} \frac{230 A_{\lambda}}{\frac{g}{100 ml\, f.a.v} \times \frac{10^{6} \mu g}{g} \times cm} = \frac{2.3 \times 10^{-2} A_{\lambda} \times cm^{-1}}{\mu g / ml\, f.a.v}$$

Whereas

$E^{1\%}$ = *absorption coefficient at the same wavelength.*

F. A. V = *final assay volume*

Results

The calibrations curves we have obtained using the reagent and the method describe above obey satisfactorily the law of Lambert and Beer.

References

1. Lowry, O.H., Rosebrough, N.J., Farr, A.L., and Randall, R.J. (1951) *J. Biol. Chem***193 265** (The original method).

2. Hartree E.E. (1972). *Anal. Biochem.* **48422** (This modification makes the assay linear over a larger range than the original assay)

3. www.ibt.lt/uploads/file/bvtl-2/assays_3.pdf

2.1.3 Bicinchinonic Acid (BCA) Assay

The bicinchoninic acid $(BCA)_2$ assays is a colorimetric method that is commonly employed to estimate the concentration of protein in a sample. In this assay, protein levels are measured via the creation of a purple, Cu+1$(BCA)_2$ chromophore (k max = 562 nm). In test tubes or 96-well plates, the relationship between protein concentration and absorbance is nearly linear over a wide working rang (0.02 to 2 1 g/l). However, assay performance is reducing in lower volume formats, a physical phenomenon common to many colorimetric assays. Despite these challenges, low-volume BCA assays are sometimes preferred, such as when the biological sample is particularly limited (e.g., patient tissues).

Principle

The bicinchinonic acid (BCA) assay for total protein is a spectophotometric assay based on the alkaline reduction of the cupric ion to the cuprous ion by the protein, followed by chelation and color development by the BCA reagent. Either a micro or a semi micro procedure, the latter generating a final assay volume of 2 or 3ml, may be used; the semi micro procedure is most frequently used. The BCA assay for total protein is somewhat variable: it has differing sensitivities in response to incubation time, incubation temperature, standard protein used for calibration, and other factors (Smith *et al.*, 1985). Certain classes of compounds such as reducing sugars and ammonium ions interfere with the assay, sometimes severely. However, if interfering compounds are eliminated (e.g., by dialysis). The BCA assay has a good combination of sensitivity and simplicity, and it has some advantages over the Lowly technique.

Materials

- Calibration standard: 1 mg BSA/ml
- Commassie blue 1x dye solution
- Buffer or solvent used to prepare the protein-containing sample
- Sample containing protein
- BCA reagent A/reagent R mix
- Spectrophotometer and cuvettes

Procedure

1. Prepare a dilution series of calibration standard in the buffer or solvent used to prepare the sample to cover the range 0.2 to 1.0 mg/ml.
2. for a 2.1-ml final assay volume: Mix 100 µl of protein-containing sample, of calibrating standard, or of buffer used to prepare the sample (reference standard) with2 ml BCA reagent N reagent B mix in separate tubes.
3. for a 4.2-ml final assay volume: Mix 200 µl of protein-containing sample, of calibrating standard, or of buffer used to prepare the sample (reference standard) with 4 ml BCA reagent A/ reagent B mix in separate test tubes. Coefficient ($E^{1\%}$) at the same wavelength. Units in the concentration factor in the denominator of $E^{1\%}$ are (dry g)/100 ml of final assay volume.

   [The final assay volume depends on the size of the available cuvettes].
4. Incubate the samples and standards 30 min at 37°C, and then cool to room temperature.
5. Measure the absorbance of the sample, calibration standards, and reference standard at 562 nm (A_{562}).If the spectrophotometer does not automatically give net absorbance readings, subtract the values for the reference standard from those obtained for the sample and calibration standards.

6. Prepare a calibration plot by graphing the net A_{650} values for the standards versus protein concentration. Determine the protein concentration of the sample by interpolation from the plot.

Results

The calibrations curves we have obtained using the reagent and the method describe above obey satisfactorily the law of Lambert and Beer.

2.1.4 Acid Digestion-Ninhydrin Assay

In this assay, proteins are hydrolyzed to amino acids with 6% sulfuric acid at 100°C. The hydrolysates are neutralized and the amino acids quantities by ninhydrin derivatization and spectrophotometry at 570 nm (A_{650}).This method is resistant to interference by Phenolic and related compounds that may affect Folin reaction dependent analyses. Individual amino acids do not produce equal "color yields" upon reaction with ninhydrin. However, most proteins-accept those with unusual compositions such as high hydroxyproline and proline (collagens), extraordinary sulfur contents (cysteine; e.g., alcohol dehydrogenase and thrionein), or densely glycosylated proteins (e.g., invertase and glycophosin) -give reasonable results using leucine as a calibration standard. Ammonium ion generates strong color with ninhydrin, nearly equivalent to that produced by leucine. However, TCA precipitation of the proteins prior to acid hydrolysis will remove NH_4^+, which will be left in the supernatant.

Materials

- H_2SO_4
- NaOH
- Sample containing protein
- Calibration standard: 0.2 mg/ml leucine, diluted ½ to ½$_0$
- Ninhydrin reagent
- Isopropanol diluent: 50% (vlv) isopropanol in water
- Heat-sealable tubes
- 100°C water bath

Procedure

1. Based on the percentage H_2SO_4 (e.g., 6%) to be used in this assay, determine the volume of NaOH required to neutralize the reaction.

   [For example, 6% H_2SO_4*is -2.3 N* (H^+)*, so it would take -15.3 ml of 6% (w/v) NaOH (1.5 N) to completely neutralize 10 ml of 6%* H_2S0_4*].*

   In this protocol, % of H_2SO_4refers to a v/v dilution of concentrated H_2SO_4. The total volume of the reaction mixture after the addition of acid base, and ninhydrin reagent to a protein solution containing 10 to 20 mg protein should be close to 3 but no more than 6 ml. If necessary, use more concentrated acid and base solutions to prevent excessive dilution of the sample.

2. In a heat-sealable **hlbe,** add H2S0, to a 0.02- to 0.05-mg protein sample to give a final concentration of 3% acid. Seal the tube under a nitrogen blanket and incubate 12 to 15 hr (Over night) at 100° to 105°C.

   [Protein in aqueous solution may be used directly it does not contain many additional amino acids and oligopeptides. If the sample is diluted or contaminated or interfering compounds, the protein can be precipitated and concentrated using TCA. After the supernatant is removed, 3% H_2SO_4*can be added directly to the TCA precipitate for hydrolysis.]*

3. Open the tube and add the appropriate volume of 6% NaOH, as determined in step 1, to neutralize the H_2SO_4, used in hydrolysis.
4. Add 2 ml ninhydrin reagent to the neutralized sample, and to equal volumes of diluted calibration standards, and of the buffer (reference standard) in separate test tubes. Incubate the mixture 20 min in 100°C (boiling) water bath. Cool to room temperature

   [Calibration and reference standards should be light blue, samples a darker purplish blue. Very dark Samples should be diluted by a factor of 2 to 5 with isopropanol diluent (dilute all standards to the same extent). Dilution may be performed after incubation with the reagent, and it is not necessary to repeat the assay in such a case.]

5. Measure the net absorbance of sample and calibration and reference standards at 570 nm (A_{570}). If the spectrophotometer does not automatically give net absorbance readings, subtract the value for the reference standard from those obtained for sample and calibration standards.

   Theabsorbance of the sample dominated by hydroxyproline and proline may be read at 440nm; a separate set of calibration standards should be prepared with these amino acids.

6. Prepare a calibration plot by plotting the adjusted values versus protein concentration (µg of protein/ ml f.a.v.). Determine the protein concentration of the sample by interpolation from the curve.

The slope of ninhydrin color yield (A570) which most amino acids are 20 to 25 A_{570} µmol amino acids/ ml f.a.v-1 cm-1, as a general guide, use 0.02 to 0.05 mg protein per assay.

Results

The calibrations curves we have obtained using the reagent and the method describe above obey satisfactorily the law of Lambert and Beer.

References

1. Christine V. S., Roger L.L. and Nicholas C. P. (1999) *Biotechnol. Appl. Biochem*. Review Colorimetric protein assay techniques **29**, 99–108.
2. www.ibt.lt/uploads/file/bvtl-2/assays_3.pdf

2.1.5 Coomassie Blue (CB) Assay

The use of the metachromatic response observed on the binding of CB to proteins for the determination of protein concentration was popularized by Bradford

(1976). The binding of CB dyes to proteins was first studied by Fazekas de St. Groth *et al.*(1963).

Principle

A protein determination method which involves the binding of Coomassie Brilliant Blue G-250 to protein is described. The binding of the dye to protein causes a shift in the absorption maximum of the dye from 365 to 595 nm. And it is the increase in absorption at 595 nm which is monitored. This assay is very reproducible and rapid with the dye binding process virtually complete in approximately 2 min with good color stability for 1 hr. There is little or no interference neither from cations such as sodium or potassium nor from carbohydrates such as sucrose. A small amount of color is developed in the presence of strongly alkaline buffering agents, but the assay may be run accurately by the use of proper buffer controls. The only components found to give excessive interfering color in the assay are relatively large amounts of detergents such as sodium dodecyl sulfate, Triton X-100, and commercial glassware detergents. Interference by small amounts of detergent may be eliminated by the use of proper controls.

Materials

- **Reagents**

  Coomassie Brilliant Blue G-250 was obtained from Sigma, and used as supplied. 2-Mercaptoethanol was obtained from Sigma. Triton X-100. Sodium dodecyl sulfate was obtained from BDH Chemicals Ltd., Poole, England. Hemosol was obtained from Scientific Products. All other reagents were of analytical grade or the best grade available.

- **Protein preparation**:

  Bovine serum albumin (2x crystallized), chymotrypsinogen A, and cytochrome **c** (horse heart). Hemoglobin and human serum albumin. Protein solutions were prepared in 0.15 M NaCl.

- **Preparation of protein reagent**

  Coomassie Brilliant Blue G-250 (100 mg) was dissolved in 50 ml 95% ethanol. To this solution 100 ml 85% (w/v) phosphoric acid were added. The resulting solution were diluted to a final volume of I liter. Final concentrations in the reagent were 0.01% (w/v) Coomassie Brilliant Blue G-250, 4.7% (w/v) ethanol, and 8.5% (w/v) phosphoric acid.

- **Protein assay (standard method):**

Protein solution containing 10 to 100 pg protein in a volume up to 0.1 ml was pipette out into 12 x 100 mm test tubes. The volume in the test tube was adjusted to 0.1 ml with appropriate buffer. Five milliliters of protein reagent was added to the test tube and the contents mixed either by inversion or vortex The absorbance at 595 nm was measured after 2 min and before I hr in 3 ml cuvettes against a reagent blank prepared from 0.1 ml of the appropriate buffer and 5 ml of protein reagent. The weight of protein was plotted against the corresponding absorbance resulting in a standard curve used to determine the protein in unknown samples.

- **Micro-protein assay:**

  Protein solution containing 1 to 10 pg protein in a volume up to 0.1 ml was pipetted into 12 x 100 mm test tubes. The volume of the test tubes was adjusted to 0.1 ml with the appropriate buffer. One milliliter of protein reagent was added to the test tube and the contents mixed as in the standard method. Absorbance at 595 nm was measured as in the standard method except in 1 ml cuvettes against a reagent blank prepared from 0.1 ml of the appropriate buffer and 1 ml of protein reagent. Standard curves were prepared and used as in the standard method.

Procedure

Protein Assay by Dye Binding

It is based on the observation that Coomassie Brilliant Blue G-250 exists in two different color forms, red and blue. The red form is converted to the blue form upon binding of the dye to protein. The protein-dye complex has a high extinction coefficient thus leading to great sensitivity in measurement of the protein. The binding of the dye to protein is a very rapid process (approximately 2 min), and the protein-dye complex remains dispersed in solution for a relatively long time (approximately1 hr), thus making the procedure very rapid and yet not requiring critical timing for the assay.

Results

Reproducibility, Sensitivity, and Linearity of the Assay

Triplicate standard assays of bovine serum albumin as a standard result in a highly reproducible response pattern. Statistical analysis gives a standard deviation of 1.2% of mean value for the assay. There is extreme sensitivity in the assay with 25 pg sample giving an absorbance change of 0.275 OD units. This corresponds to 5 pg protein/ml in the final assay volume. There is a slight nonlinearity in the response pattern. The source of the nonlinearity is in the reagent itself since there is an overlap in the spectrum of the two different color forms of the dye.

2.1.6 Silver Staining of Proteins

Thefractionated polypeptides on gels are visualized, by staining generally either with coomassic blue R 250 or amido black 10B dye. The above dye can be detecting a band containing little as 0.1 μg of polypeptide. In many occasions, the available protein for electrophoresis is so small or some proteins occurs in a minute amounts the detection becomes extremely difficult with these dyes. Under such a circumstance a higher sensitive detection system is required. Silver staining is a very useful method in this regard with about autoradiography of labeled polypeptides. There are different methods described by different workers for silver staining. The method given below very is simple and rapid.

Principle

In silver staining, the gel is Impregnated with soluble silver ions and developed by treatment with formaldehyde, which reduces silver ions to form an insoluble brown precipitate of metallic silver. This reduction is promoted by protein.

Materials

- **Fixative (200 ml):**

  Mix 100 ml of methanol with 100 ml of water and 100µl of formaldehyde. Formaldehyde is added freshly
- **Dithiothreitol (DTT) (1 mg/ ml):**

  Dissolve 10 mg of DTT in 10 ml of water. Store as 1ml aliquots at −20°C
- 20% Silver Nitrate (25 ml):

  Dissolve 5 g of silver nitrate in 25 ml of water. Store in an amber-colored bottle at 4°C
- **4% Sodium Carbonate, freshly prepared (500 ml):**

  20 g of sodium carbonate dissolved in 500 ml of water. Add 400µl of formaldehyde to it
- **2.5M Citric acid (25 ml):**

  Dissolve 13.13 g of citric acid in 25 ml of water
- **Gel tray**
- **Gel shaker**

Procedure

1. Transfer the gel to a tray containing fixative solution.
2. Keep it covered in a shaker for 2 hr under cover.
3. The size of the gel shrinks, indicating proper fixation.
4. Rinse the gel once with distilled water.
5. Add 200 ml of water and two aliquots of DTT to get a final concentration of 10µg/ ml DTT.
6. Incubate the gel by shaking 30–45 min. This swells the gel back to its original size.
7. Rinse once with 200 ml of water.
8. Add 200 ml of water and 1.5ml of 20% silver nitrate.
9. Incubate by shaking for 30–45 min.
10. Wash thrice with 200 ml of water to completely remove the silver nitrate.
11. Pre-rinse with 4% sodium carbonate containing formaldehyde.
12. Allow the gel to develop with a fresh solution of sodium carbonate.
13. Stop the reaction with 2.5M citric acid after the development of bands.
14. Wash the gel finally in water.

Precautions

- Water used should be deionized water only
- Do not touch the gel with bare hands. Use gloves.
- All solutions should be at RT before use.

2.1.7 Western Blotting

Western Blotting or immunoblotting allows determining the presence of a specific protein in a sample after separation on SDS-PAGE. The term Western blotting is used after a similar term 'Southern blotting', which was invented by and named after E. M. Southern. That method allows for detection of nucleic acids in a blot technique. The proteins are separated on SDS-PAGE and then transferred to a membrane (generally nitrocellulose or PVDF). The membrane is incubated with a source of non-specific protein (such as milk proteins) to bind to any remaining sticky places on the membrane. A primary antibody is then added to the solution which is able to bind to its specific target protein followed by washes and incubation in a solution of secondary antibody. The secondary antibody recognizes the primary antibody and binds at locations on the blot where the primary antibody is bound as well. The secondary antibody is furthermore conjugated with an enzyme or marked with a nucleotide, thus allowing detection.

Principle

The separated proteins are transported out of the gel either by the capillary action of buffer or in an electric field. The presence of SDS increases the solubility of proteins and thus facilitates the migration of proteins. Once out of the gel, the proteins comes in contact with the nitrocellulose membrane which binds the protein very strongly onto to surface as a thin band thus producing a replica of the original gel.

Materials

- Transfer buffer (1 litre): Dissolve 5.82 g of Tris base, 2.93 g of Glycine, 3.75 ml of 10% SDS in 800 ml water. Make it to 1 litre with 200 ml of Methanol
- TBST (1 litre): Dissolve 8.8 g of Tris base, 1.2 g of Sodium Chloride and 500µl of Tween 20 in 750 ml of water. Adjust the pH of the solution to 7.5 and finally bring the volume to 1 litre
- Blocking solution: 5 g of fat free protein in 100 ml of TBST.
- Stripping buffer (1 litre): To 125 ml of 0.5M Tris (pH 6.8) buffer add 100 ml of 10% SDS and 8ml of 2-Mercaptoethanol. Bring the volume to 1 litre with water.

Procedure

1. After the gel run, transfer into the transfer buffer.
2. Soak the filter paper in the transfer buffer and lay flat.
3. Soak the nitrocellulose membrane and lay it on top of the filter paper.
4. Place the gel on top of the nitrocellulose membrane.
5. Layer the gel again with a filer paper soaked with transfer buffer
6. Roll a pipette so as to remove trapped air bubbles.
7. Transfer the sandwich directly to a transfer apparatus in case of semi-dry apparatus, or to a cast in case of wet transfer.
8. Place the gels with the membrane on the anode or positive electrode

(usually Red) and the gel on the cathode or negative electrode (usually Black).

9. Cover the apparatus and transfer as per the specification of the instrument as stated in the manufacturer's instructions.
10. After the transfer, mark the blot and then wash with TBST buffer.
11. Block the membrane in blocking solution for at least 1 hr at RT or overnight at 4°C.
12. After blocking, incubate the blot with primary antibody at an appropriate dilution in 10ml of blocking solution.
13. Incubate on rocking, shake for at least 2 hr at RT or overnight at 4°C.
14. Wash the blot 3 times for 5 min each with TBST.
15. Add 15 ml of blocking solution with appropriate amount of secondary antibody (Either Alkaline phosphatase or HRP conjugated antibody).
16. Incubate the blot on the rocker for 2 hr at RT.
17. Wash the blot 3 times for 5 min each with TBST.
18. Develop the blot with the respective developer depending on the secondary antibody.

Blot Development

When using Alkaline Phosphatase based secondary antibody:

- Add the commercially available substrate solution containing 5-Bromo-4-chloro-3-indolyl Phosphate (BCIP)/ Nitroblue Tetrazolium (NBT) (SIGMA) to the blot till it is just submerged.
- Incubate the blot in the substrate solution for 10–30 minutes until the development of bluish purple color.
- Stop the reaction by washing the strips in several changes of water.
- Air-dry the strip and photograph or scan it for records.

When using the HRP conjugated secondary antibody:

- Just before developing, prepare the substrate solution by mixing equal parts of reagents 1 and 2 (Commercially available like Pierce ECL Western Blotting Substrate).
- Incubate the blot with gentle shaking in the substrate for 1min at RT.
- Using a forceps, lift the blot from the substrate and drain off the excess solution by placing the tip of the blot on a filter paper.
- Wrap the blot in a saran wrap.
- Place the blot in a film cassette.
- Take the cassette to the developing room without lights except for the safe light, place an X-ray film on the top of the blot.
- Expose the film to the blot for various time points.
- Develop the film.

Reprobing blot

- Wash the blot once with TBST.
- Add 10 ml of stripping buffer to the blot in an air tight container.
- Strip the blot at 50°C for 30 min.
- Wash the blot till no smell of 2-Mercaptoethanol can be detected.
- Block the blot and use for reprobing.

2.2 Total Protein Estimation by Lowry's Method

Introduction

Protein can be estimated by different methods as described by Lowry and also by estimating the total nitrogen content. No method is 100% sensitive. Hydrolyzing the protein and estimating the amino acid alone will give the exact quantification. The method developed by the Lowry *et. al* (1951) is sensitive enough to give a moderately constant value and hence largely followed. Protein content of enzyme extracts is usually determined by this method.

Principle

The phenolic group of tyrosine and trytophan residues (amino acid) in a protein will produce a blue purple color complex, with maximum absorption in the region of 660 nm wavelength, with Folin- Ciocalteau reagent which consists of sodium tungstate molybdate and phosphate. Thus the intensity of color depends on the amount of these aromatic amino acids present and will thus vary for different proteins. Most proteins estimation techniques use Bovin Serum Albumin (BSA) universally as a standard protein, because of its low cost, high purity and ready availability. The method is sensitive down to about 10µg/ ml and is probably the most widely used protein assay despite its being only a relative method, subject to interference from Tris buffer, EDTA, nonionic and cationic detergents, carbohydrate, lipids and some salts. The incubation time is very critical for a reproducible assay. The reaction is also dependent on pH and a working range of pH 9 to 10.5 is essential.

Materials

1. BSA stock solution (1mg/ml),
2. Analytical reagents:
   a) 50 ml of 2% sodium carbonate mixed with 50 ml of 0.1 N NaOH solution (0.4 gm in 100 ml distilled water.)
   b) 10 ml of 1.56% copper sulphate solution mixed with 10 ml of 2.37% sodium potassium tartarate solution. Prepare analytical reagents by mixing 2 ml of (b) with 100 ml of (a)
3. **Folin - Ciocalteau reagent** solution (1N) Dilute commercial reagent (2N) with an equal volume of water on the day of use (2 ml of commercial reagent + 2 ml distilled water)

4. **Preparation of Standard for Proteins (Stock)**
   a) Weigh accurately 50 mg of bovine serum albumin and dissolve in distilled water and make up to 50 ml in a standard flask.
   b) Working standard:

      Dilute 10 ml of the stock solution to 50 ml with distilled water in a standard flask. One ml of this solution contains 200 µg protein.

Procedure

- Preparation of standards:
  1. Different dilutions of BSA solutions are prepared by mixing stock BSA solution (1 mg/ ml) and water in the test tube as given in the table. The final volume in each of the test tubes is 5 ml. The BSA range is 0.05 to 1 mg/ ml.
  2. From these different dilutions, pipette out 0.2 ml protein solution to different test tubes and add 2 ml of alkaline copper sulphate reagent (analytical reagent). Mix the solutions well.
  3. This solution is incubated at room temperature for 10 mins.
  4. Then add 0.2 ml of reagent Folin Ciocalteau solution (reagent solutions) to each tube and incubate for 30 min. Zero the colorimeter with blank and take the optical density (measure the absorbance) at 660 nm.
  5. Plot the absorbance against protein concentration to get a standard calibration curve.
  6. Check the absorbance of unknown sample and determine the concentration of the unknown sample using the standard curve plotted above.

Extraction of Protein from Sample

- Extraction is usually carried out with buffers used for the enzyme assay. Weigh 500 mg of the sample and grind well with a pestle and mortar in 5 – 10 ml of the buffer. Centrifuge and use the supernatant for protein estimation.

Estimation of Protein from Sample

1. Pipette out 0.2, 0.4, 0.6, 0.8 and 1 ml of the working standard into a series of test tubes.
2. Pipette out 0.1 ml and 0.2 ml of the sample extract in two other test tubes.
3. Make up the volume to 1 ml in all the test tubes. A tube with 1 ml of water serves as the blank.
4. Add 5ml of reagent alkaline copper solution to each tube including the blank. Mix well and allow to stand for 10 min.
5. Then add 0.5ml of reagent *Folin Ciocalteau Reagent,* mix well and incubate at room temperature in the dark for 30 min. *Blue color is developed.*

6. Take the readings at 660 nm.
7. Draw a standard graph and calculate the amount of protein in the sample.

Observations

BSA (ml)	Water (ml)	Sample conc. (mg/ml)	Sample vol. (ml)	Alk. CuSO4 (ml)	Lowery reagent (ml)	Sample (ml)	O.D. 660 nm
0.25	4.75	0.05	0.2	2	0.2	2	
0.5	4.5	0.1	0.2	2	0.2	2	
1	4	0.2	0.2	2	0.2	2	
2	3	0.4	0.2	2	0.2	2	
3	2	0.6	0.2	2	0.2	2	
4	1	0.8	0.2	2	0.2	2	
5	0	1.0	0.2	2	0.2	2	

2.3 Estimation of Free Amino Acids

2.3.1 Extraction and Quantifications Procedures of Amino Acids

Amino acids are the monomers from which proteins are made. All amino acids contain a primary amine group and a carboxylic acid group, together with other groups that determine the differences in their chemical properties. Amino acids can undergo further modification after they have been incorporated into cellular proteins. Analysis of amino acids requires careful separation by chromatographic methods. Amino acids are an important fraction in most foodstuffs, ranging from 1.8% in roots and tubers up to about 25% in legumes. However, in foods they are present in a complex multicomponent, and some minor components, all of which might interfere during the analysis of proteins in foods. A whole variety of analytical tools for amino acid analysis are available to the analyst but not all of them are suitable for routine analysis procedures. This article will deal with some key analytical techniques in order to give an overview of possibilities and shortcomings of current techniques and some promising future trends. Well established methods are the spectroscopy in the ultraviolet (UV), visible (VIS), and infrared (IR) ranges of the spectrum, chromatography, and conventional electrop Horesis. Since validation of methods and results are important features of today's analysis, some attention is paid to the availability and use of reference materials.

H
|
R—C—CO_2H
|
NH_2

Basic Structure of the Amino Acid

The common feature of the structure of amino acids are having a minimum of two ionizable groups; the acidic carboxylic group (-COOH COO^- + H^+) and the basic amino group (-NH_2^+ H^+ NH_3^+) on the same carbon atom, called alpha carbon atom. Because of this the amino acids undergoes reactions characteristic of these groups.

2.3.2 Quantitative Tests for Amino Acids

There are a number of quantitative tests to detect the presences of amino acids and these largely depend on the nature of R- group and hence specific. These methods are used sparingly in recent days as they have no general value and hence are only mentioned briefly. The following test can be carried out with the number of amino acids (1% concentration) and specificity checked.

Sr. No.	Experiments (Tests)	Observations	Conclusion
1.	**Xanthoproteic Reaction** Addfew drops of Conc. HNO_3 to tissue	Tissue becomes yellow	The Phenolic groups of tyrosine unit of the protein react with Conc. HNO_3, sample turns to yellow
2.	**Millon's Reactions** Prepare a 15% solution of mercuric sulphate in 15% H_2SO_4. Add a few drops of this reagents to a 1 ml of the test solution and heat it for 10 minutes, at 100°C. After cooling add few drops of 1% sodium nitrate solution.	Test sample becomes Red in color	The aromatic compounds such as tyrosine react with Millon's reagent to form a red colored complex.
3.	**Hopkins-cole Test** Thereagent is glyxolic acid which is obtaining by exposing glacial acetic acid to sunlight for few minutes. To 2ml of this solution, add 2 ml of the test solution mix well and carefully add to the side of the tube 2ml of conc. H_2SO_4.	A violet ring at the junction indicates the presences of tryptophan	The indole group of tryptophan reacts with glyoxylic acid. In the presence of concentrated H_2SO_4 to give a purple color.
4.	**Ehrlich's Test** The reagent consists of 10% *p*-dimethyl amino benzaldehyde in 10 % HCl. To 1 ml of reagent, 1 ml of sample is added. This test is specific for indole which gives red color.	Sample in the test tube is red in color.	Aromatic amines and many organic compounds (indole and urea) give a colored complex with this test. Apply this test to tryptophan, urea and glycine.
5.	**Pauly's test** Mix 1 ml of sulphonic acid (1% solution in 10%HCl) with 2 ml of test solution and cool in ice. Add 1ml of 5% sodium nitrite solution. After 5 minutes add 2 ml of 1% Na_2CO_3 solution and note change in color.	If Color is develop	Sulphanilic acid upon diazotization in the presence of sodium nitrite and hydrochloric acid results in the formation a diazonium salt. The diazonium salt formed, couples with either tyrosine or histidine in alkaline medium to give a red colored chromogen (azo dye).

6.	Lead Sulphide Test: The test solution (2ml) boiled with 1N NaOH (0.5 ml) for 2 minutes and cooled. The reagent sodium plumbate is prepared by mixing 5 ml of 0. 1 N NaOH with 2 ml of 0.1 M lead acetate and heated till the white precipitated dissolve. About 0.5 ml of this reagent is added to this test solution.	Formation of black precipitates indicates the presence of sulphur containing amino acids.	Sulphur containing amino acids, such as cysteine and cystine upon boiling with sodium hydroxide (hot alkali) yield sodium sulphide. This reaction is due to partial conversion of the organic sulphur to inorganic sulphide, which can detect by precipitating it to lead sulphide, using lead acetate solution.
7.	Sakaguchi Test: 3 ml of the test solution is mixed with 1 ml of 1 N NaOH solution and 2 drops of alpha napthol (1% alcoholic solution) is added to this mixture. A few drops of bromine water (add few drops of bromine in 100 ml of water) are added.	Note the red color.	Under alkaline condition, α- naphthol (1-hydroxy naphthalene) reacts with a mono – substituted guanidine compound like arginine, which upon treatment with hypobromite or hypochlorite, produces a characteristic red color.

2.4 Estimation of Total Free Amino Acids

The amino acids are colorless ionic compounds that form the basic building blocks of proteins. Apart from the bring bound as proteins, amino acids also exist in the free from in many tissue and are well known as free amino acids. They are mostly water soluble in nature. Very often in plants during disease conditions, the free amino acid composition exhibits a changes and hence, the measurement of the total free amino acids gives the physiological and health status of the plants.

Principle

Ninhydrin, a powerful oxidizing agent, de carboxylates the alpha- amino acids and yields an intensely colored blush purple product which is colorimetrically measured at 570 nm.

Ninhydrin + alpha- amino acid → Hydrindantin + Decarboxylated Amino acid + Carbon dioxide+Ammonia.

Hydrindantin + Ninhydrin + Ammonia → Purple colored product + water

Materials

- Ninhydrin: dissolve 0.8g stannous chloride ($SnCl_2$. $2H_2O$) in 500 ml of 0.2 M citrate buffer (*pH 5.0*). Add this solution to 20g of Ninhydrin in 500 ml of methyl cellosolve (2 methoxy ethanol).
- 0.2 Citrate Buffer pH 5.0
- Dilute Solvent: Mix equal volumes of water and *n*- propanol, and use.

Procedure

Extraction of Amino acids

Weigh 500 mg of the plant sample and grind it in a pestle and mortar with a small quantity of acid- washed sand. To this homogenate, add 5 to 10 ml of 80% ethanol. Filter or centrifuge. Save the filtrate or the supernatants. Reduce the volume if needed by evaporation and use the extract for the quantitative estimation of total free amino acids. If the tissue is tough, use boiling 80% ethanol for extraction.

Estimation

1. To 0.1 ml of extract, add 1 ml of Ninhydrin solution.
2. Make up the volume to 2 ml with distilled water
3. Heat the tube in a boiling water bath for 20 min.
4. Add 5 ml of the diluent and mix the contents
5. After 15 min read the intensity of the purple color against a reagent blank in a colorimetric at 570 nm. The color is stable for 1 hour.
6. Prepare the reagent blank as above as by taking 0.1 ml of 80% ethanol instead of the extract.

Standard

Dissolve 50 mg leucine in 50 ml of distilled water in volumetric flask. Take 10 ml of this stock standard and dilute to 100 ml in another volumetric flask for working standard solution. A series of volume 0.1 to 1 ml of this standard solution gives a concentration range 10 μg to 100 he μg. proceed as that of sample and read the color.

Results

Draw a standard curve using absorbance versus concentration. Find out the concentration of the total free amino acids in the sample and express as percentage equivalent of leucine.

Notes

1. Ninhydrin is carcinogenic, wear gloves while handling it.
2. Since this estimation includes only alpha-amino acids, non amino acids are not counted.

2.5 Estimation of Proline

Introduction

Proline is basic amino acid found in high percentage in a basic protein. It play role as anti tress in plant. Many researchers have reported a several-fold increase in the proline content under physiological and pathological stress conditions. Hence, the analysis of proline in plants has become in pathology and physiology.

Principle

During selective extraction with aqueous sulphosalicylic acid, proteins are precipitate as complex, other interfering materials are also presumably removed by absorption to the protein sulphosalicylic acid complex. The extracted proline is made to react with ninhydrin in acidic conditions (PH 1.0) to from the chromophore (red color) and read at 520 nm.

Materials

- Acidic ninhydrin:
- Warm 1.25 g ninhydrin in 30 ml glacial acetic acid and 20 ml 6 M pHospHoric acid, with agitation until dissolved. Store at 4°C and use within 24hrs.
- 3% of aqueous Sulphosalicylic acid
- Glacial acetic acid
- Toluene
- Proline

Procedure

1. Extract 0.5 g of plant material by homogenizing in 10 ml of 3% aqueous sulphosalicylic acid.
2. Filtrate the homogenate through Whatman No.2 filter paper
3. Take 2ml of filtrate in a test tube and add 2ml of glacial acetic acid and 2 ml of ninhydrin.
4. Heat it in the boiling water bath for 1 hr.
5. Terminate the reaction by placing the tube in ice bath.
6. Add 4ml toluene to the reaction mixture and stir well for 20 -30 sec.
7. Seprate the toluene layer and warm to room temperature.
8. Measure the red color intensity at 520 nm
9. Run a series of standard with pure proline in a similar way and prepare a standard curve.
10. Find out the amount of proline in the test sample from the standard curve.

Calculations

Calculate the proline content on fresh-weight-basis as follows:

$$\mu\text{moles per g tissue} = \frac{\mu\text{g proline/ ml} \times \text{ml toluene}}{115.5} = \frac{5}{\text{g of sample}}$$

Where 115.5 is the molecular weight of proline

Notes

1. The color intensity is stable for at least 1hr.
2. The relationship between the amino acid concentration and absorbance is linear in the range 0.02 to 0.1 μM per ml of proline.

2.6 Estimation of Lysine

Introduction

Lysine is a limiting amino acid in the cereal grains. Assessment of lysine in cereal grains for nutritional quality is hence one of the procedures adopted for screening varieties. Though an ideal sample method for direct estimation of lysine is yet to be found, the given procedure is sufficient for routine screening.

Principle

The protein in the grain sample is hydrolysed with a proteolytic enzyme, papain. The alpha amino group of lysine which does not couple with copper is made to form alpha- dinitropyridyl derivative of lysine with 2- chloro-3, 5-dinitropyridine. The excess pyridine is removed with ethyl acetate and the color of alpha-dinitropyridyl derivative is read at 390 nm.

Materials

- Solution A: 2.89 $CuCl_2$. $2H_2O$ in 100 ml of water.
- Solution B: 13.6g Na_2 PO_4, $12H_2O$ in 200 ml water.
- Sodium borate buffer: 0.05 M pH 9.0
- Copper phosphate reagent: pour solution a (100 ml) to B (200 ml) with swirling, centrifuge and discard the supernatant. Re-suspend the pellet three times in 15 ml borate buffer and centrifuge after each suspension. The reagent must be prepared fresh for every seven days.
- 3% solution of 2-Chloro-3, 5-Dinitropyridine in methanol. Prepare fresh just prior to use.
- 0.05 M Sodium carbonate Buffer pH 9.0
- Amino acid mixture: grind in a mortar 30 g alanin, 50 mg glutamic aicd, 60 mg aspartic acid, 20mg isoleucine, 80mg leucine, 30mg methionine, 40mg phenylanine, 80mg proline, 50 mg serine, 30mg threonine, 30mg tyrosine, and 40mg valine. Dissolve 100 mg of this mixture in 10 ml of sodium carbonate buffer (0.05M pH 7.0)
- 1.2 N HCl
- Ethyl acetate

Procedure

1. To 100mg of defatted grain sample add 5 ml of papain solution and incubate overnight at 65°C. Cool to room temperature, centrifuge and decant the clear digest.
2. To one ml digest taken in acentrifuge tube add 0.5 ml carbonated buffer and 0.5 ml copper phosphate suspension.
3. Shake the mixture for 5 min in a vortex mix and centrifuge.
4. To one ml of supernatant add 0.1 ml pyridine reagent, mix well and shake for 2 hr.
5. Add 5 ml of 1.2 N HCl and mix

6. Extract three times with 5 ml ethyl acetate and discard the ethyl acetate (top) layer.
7. Read the absorbance of aqueous layer at 390 nm
8. Prepare a blank with 5 ml papain alone repeating steps 1 to 7.
9. Dissolved 62.5 mg lysine monohydrochloride in 50 ml carbonated buffer (1mg lysine/ml). Pipette out 0.2 0, 0.4, 0.6, 0.8, and 1 ml and make up with one ml carbonated buffer. Add 4 ml papain to each tube and mix 0.5 ml of copper phosphate suspension. Carry out steps 3 to 7. The standard curve represents absorbance values for 40, 80, 120, 160 and 200 μg lysine.

Calculations

Prepare a standard curve from the readings of the standard lysine. Subtract the absorbance of the blank from that of the sample and calculate the lysine content in the aliquot from the graph.

$$\text{Lysine content of the sample} = \frac{\text{Lysine value from graph in } \mu g \times 0.16}{\text{Percent N in the sample}}$$

$$= \text{g per 16 g N.}$$

2.7 Estimation of Methonine

Introduction

Methonine is one of the essential, sulphur containing amino acids. Although it is present in many food proteins, methionine estimation is a routine in nutrition screening programmers of grain legumes.

Principle

The proteins in the grain are first hydrolysed under mild acidic condition. The liberated methionine gives a yellow color with nitroprusside solution under alkaline condition and turns red on acidification. The glycine is added to the reaction mixture in order to inhibit color formation with other amino acids.

Materials

- 2 N Hydrochloric acid
- 10N NaOH
- 10% NaOH
- 3% Glycine
- Orthophosphoric Acid
- HCl and dilute with water to 100ml

Procedure

1. Weigh 0.5 g of defatted sample into a 50 ml conical flask. Add 6ml of 2N HCl and autoclave at 15lb pressure for one hour.
2. Add a pinch of activated charcoal to the hydrolysate (autoclaved sample) and heat to boil. Filter when hot and wash the charcoal with hot water.

3. Neutralize the filter with 10 N NaOH to pH 6.5. Make up the volume to 50 ml with water after cooling to ambient temperature.
4. Transfer 25 ml of the made up solution into a 100 ml conical flask.
5. Add 3ml of 10% NaOH followed by 0.15 ml sodium nitroprusside
6. After 10 min add 1 ml of glycine solution.
7. After another 10 min add 2 ml ortho-phosphoric acid and shake vigorously.
8. Read the intensity of red color after 10 min a t 520nm against blank prepare in the same way but without nitroprusside.
9. Standard curve: pipettc out 0,1, 2, 3, 4 and 5ml of standard methionine solution and make up to 25 ml with water. Follow step 5 to 8 to develop the color in the standards. The level serves as the blank.

Calculation

Draw a standard curve and calculate the methionine content from the graph. Methionine content of the sample

$$= \frac{\text{Methionine content from the grapH} \times 6.4}{\text{Percentage of N in the sample}}$$

$$= \text{g per 16g N}$$

2.8 Estimation of Tryptophan

Introduction

Number of cereals like Maize and sorghum are deficient in tryptophan in additional to lysine; Methods to determine these essential amino acids are useful to identify nutritional superior types.

Principle

The indole ring of tryptophan gives an orange-red color with ferric chloride under strongly acidic condition. The color intensity is measured at 545nm.

Materials

- Papain Solution: dissolve 400mg technical grade papain in 100 ml sodium acetate buffer pH 7.0. Prepare the solution fresh every day.
- Reagent A: Dissolve 135 mg FeCl3. $6H_2O$ in 0.25ml water and dilute to 500 ml with glacial acetic acid containing 2% acetic anhydride.
- Reagent B: 30 N H_2SO_4.
- Reagent C: Mix equal volumes of reagents A and B about one hour before use.
- Standard tryptophan: dissolve 5mg tryptophan in 100ml water (50 μg/ml)

Procedure

1. Weigh100 mg of air-dried dried powdered and defatted grain sample into a small test tube.

2. Add5ml papain solution shake well and close the tube
3. Incubate at 65°C Overnight.
4. Cool the digest to room temperature, centrifuge and collect the clear supernatant.
5. Add 4 ml of reagent C in to a 1 ml of supernatant.
6. Mix in a vortex mixer and incubate at 65°C for 15 min.
7. Cool to room temperature and read the orange-red color at 545nm.
8. Set a blank with 5ml papain alone and repeat step 3- 7.
9. Pipette out 0, 0.2, 0.4, 0.6, 0.8 and 1 ml standard tryptophan and make up to 1 ml with water. Develop color following the step 5 to 7.

Calculations

Draw a standard curve, subtract the absorbance value of the blank from that of the sample and calculate tryptophan content from the graph.

Tryptophan in grain sample

$$\frac{\text{Tryptophan value from graph in } \mu g \times 0.096}{\text{Percent N in the sample}}$$

$$= g/16\ Gn$$

2.9 Estimation of Total Amino Acid by Spectro-photometric Methods

Sample Preparation for Amino Acid Estimation

Principle

For the estimation of amino acid composition of foodstuff, feed or any protein, it has to be first hydrolyzed. The free amino acids may be extracted from the tissue or food in the ethanol. The amino acid may be determined by chromatograpHy techniques as described above. If the extracted sample contains higher amount of salt and/ or sugars they have to remove before estimation of amino acids.

A. Sample preparation for free amino acids

Materials

- Ethanol
- 0.01N HCl

Procedure

1. Weigh accurately sufficient quantify of the sample which should have 2 to 6 μ mole of each amino acids
2. Extract with warm (60°C) 70% ethanol or 0.01 M phosphate buffer pH 7.0, three to six times. Use extractant five times the weight of sample for each extract.
3. Pool the extracts after filtration or centrifugation and evaporate in a rotary vacuum evaporator to dryness.

4. Take the residue in 1- 10 ml of 0.01 N HCl or in a suitable sample diluting buffer.

B. Acid Hydrolysis of Proteins

Materials

- 6 N HCl
- 0.01 N HCl

Procedure

1. Weigh accurately sufficient quantity of sample containing about 10 mg protein in a thick- walled heavy glass tube having a constriction at 7.0 cm from top
2. Add 5 ml of 6 N HCl and place the tubes in liquid nitrogen
3. Evacuate the frozen sample to 0.01mm Hg.
4. Thaw the sample by shaking and slight warming to allow any dissolved air to bubble out.
5. Freeze the contents and evacuate to 0.005 mm Hg again.
6. Seal the tube at the constriction using a flame till the tube is under evacuation.
7. Place the tube in a hot air oven at 110 $\pm$1°C for 22h to hydrolyse proteins.
8. Break open the seal and evaporate the contents in a rotary evaporator to remove hydrochloric acid.
9. Add water and repeat evaporation two times.
10. Take the residue in 1- 10 ml 0.01 N HCl or 50 mm citrate buffer (pH 2.2).

Note

Some amino acids may be degraded partly or completely during hydrolysis. Glutamine and aspargine are converted to their respective acids. Cystine may be converted to cysteic acid. Serine and threonine are very rapidly hydrolysed from the protein and are lost to some extent. Tryptophan is also partially destroyed during hydrolysis.

C. Deproteinizaction

Materials

- 1% Picric acid
- Dowex 1x 8 Cl^- Resin
- 0.02 N HCl
- 0.01 N HCl

Procedure

1. Add 50 ml of 1% picric acid to 10 ml of the sample
2. Centrifuge at 3000 rpm for 10 min and collect the supernatant.

3. Place the supernatant in a glass column (2 x 10 cm) containing Dowex 1 x 8 Cl- resin to height of 3cm
4. Immediately after the sample sinks into the resin wash the sides of the column with the elute and evaporate it to dryness
5. Take the residue in 0.01N HCl.

Note

Centrifugation of the sample for 30 minute at 2,000g is a better method for deproteinization. The ultrafiltration may be used for analysis. The picric acid does remove some amino acids.

D. Desalting

Material

- Dower 5 x 8 (H^+) or Amberlite CG- 120 (Na^+)
- 2 N NH_4OH
- 0.01N HCl

Procedure

1. Load 10 – 20 ml sample over the column of reisn (2 x 10 cm) and allow it to be absorbed by the resin
2. Elute the amino acids with ammonium hydroxide.
3. Collect the elute about 100 ml and evaporation to remove ammonium
4. Wash the residue twice with water and repeat evaporation.
5. Take the amino acid residue in 0.01 N HCl.

Note

This procedure removes sugar impurities.

E. Hydrolysis of Larger Sample

The feeds and foods being highly heterogenous, it is advisable to take larger samples to get better result.

Procedure

1. Weigh 2 g sample into a 2- liter round bottom flask
2. Add 900 ml of 6 N HCl
3. Pass nitrogen gas (oxygen –free) for 30 minutes.
4. Reflux for 20h in an oilbath at 137°C with continuous passage nitrogen
5. Filter through suction and dilute to 1 litre distilled water
6. Evaporate an aliquot of it (containing about 25 mg protein) in a rotary evaporator.
7. Wash with water twice and evaporate to dryness
8. Take the residue in 0.01 N HCl for analysis.

2.10 Extraction and Quantification of Amino Acids by HPLC Technique

Introduction

The continuous improvement in HPLC columns and instrumentation presents an opportunity to improve HPLC methods. A proven ortho- phthalaldehyde /9-fluorenylmethyl chloroformate (OPA/FMOC) derivatized amino acid analysis method developed on the HP 1090 Series HPLC Systems, and later updated for the Series HPLC Systems, has now evolved further taking advantage of the Agilent 1200 Series SL and Agilent ZORBAX Eclipse Plus C18 stationary phase columns.

A chromatogram is presented for each column or method.

They are grouped into three categories:

- Traditional 5-µm particle size columns that offer high resolution
- Practical 3.5-µm particle size columns that combine high resolution and speed of the 5 and 1.8-µm particle size columns (Rapid Resolution)
- Highest throughput 1.8-µm particle size columns that offer (Rapid Resolution High Throughput) excellent resolution.

The 2.1 × 150 mm, 3.5 µm Rapid Resolution method is a preferred replacement method for the original AminoQuant method that used column p/n 79916AA-572.

All of the methods use the same chemicals and OPA/FMOC derivatization based on previous Agilent methods. The differences among these methods are the mobile phase gradient programs and the flow path. These two parameters are optimized according to column dimensions and LC instrument models.

Sample Preparation

Amino acid standard mixtures and internal standards were obtained from Phenomenex and combined into a single standard mix for analysis. The standard mixtures contained the following 36 amino acids, each with a starting concentration of 200 nmol/mL.

Glutamine (GLN), Arginine (ARG), Serine (SER), Citrulline (CIT), 1-Methylhistidine (1-MHIS), 3-Methylhistidine (3-MHIS), Glycine (GLY), 4-hydroxyproline (HYP), Glycine-proline dipeptide (GPR), Sarcosine (SAR), Alanine (ALA), Gamma-amino-butyric acid (GABA), Beta amino-isobutyric acid (BAIB), Alpha-aminobutyric acid (ABA), Proline (PRO), Methionine (MET), Ornithine (ORN), Valine (VAL), Aspartic acid (ASP), Histidine (HIS), Glutamic acid (GLU), Lysine (LYS), TryptopHan (TRP), Leucine (LEU), Isoleucine (ILE), Phenylalanine (PHE),Aminopimelic acid (APA), Cystathionine (CTH), Cystine (C-C), Tyrosine (TYR), Alpha-aminoadipic acid (AAA), Threonine (THR), Asparagine (ASN), Delta-hydroxylysine (HLY), prolinehydroxyproline dipeptide (PHP), Thiaproline (TPR).

The internal standard mix contained the following 3 amino acids each at a starting concentration of 200 n mol/ml. Homo-arginine (HARG), d3-methionine(d3-MET), homo- phenylalanine(HPHE) Sample preparation consisted of a fast SPE step (ion exchange in packed pipette tips) followed by derivatization with propyl chloroformate and subsequent clean-up by liquid-liquid extraction. The derivatization protocol resulted in amino acid species with greater chromatographic retention, resulting in improved separation of polar amino acids. The derivatized amino acids may also show improved ionization in +ESI due to blocking of acidic functional groups.

Chemical Preparation

Mobile Phase A

10 mm Na_2HPO_4: 10 mm $Na_2B_4O_7$, pH 8.2: 5 mm NaN_3F or 1 liter: (1.4 g anhydrous Na_2HPO_4 + 3.8 g $Na_2B_4O_7{\cdot}10H_2O$ in 1 L water + 32 mg NaN_3). Adjust to about pH 8.4 with 1.2 ml concentrated HCl, and then add small drops until pH 8.2. Allow stirring time for complete dissolution of borate crystals before adjusting pH. Filter through 0.45-μm regenerated cellulose membranes.

Mobile Phase B

Acetonitrile: Methanol: Water (45:45:10, v: v: v).

All mobile-phase solvents are HPLC grade. Since mobile phase A is consumed at a faster rate than B, it is convenient to make 2 L of A for every 1 L of B.

Injection Diluent

100 ml of mobile phase A + 0.4 mL concentrated H_3PO_4 in a 100-mL bottle. Store at 4°C.

0.1N HCl

Add 4.2 mL of concentrated HCl (36 %) to a 500-mL volumetric flask that is partially filled with water. Mix, and fill to mark with water. Solution is for making Extended Amino Acid and Internal Standard stock solutions. Store at 4°C.

Derivatization Reagents

Borate buffers, OPA and FMOC, are ready-made solutions supplied by Agilent. They simply need to be transferred from their container into an autosampler vial. Some precautions include:

OPA is shipped in ampoules under inert gas to prevent oxidation. Once opened, the OPA is potent for about 7-10 days. It is recommended to transfer 100-μL aliquots of OPA to microvial inserts, label with name and date, cap, and refrigerate. Replace the OPA auto-sampler micro vial daily. Each ampoule lasts 10 days.

- FMOC is stable in dry air but deteriorates in moisture. It is also recommended to transfer 100-μL aliquots of FMOC to micro vial inserts, label with name and date, cap tightly, and refrigerate. Like the OPA, an open FMOC ampoule transferred to 10 micro vial inserts should last 10 days (one vial/day).
- Borate buffer can be transferred to a 1.5-mL autosampler vial without a vial insert. It can be replaced every 10 days.

Preparing the Amino Acid Standards

Amino Acid Standards (10 pmol/μL to 1 nmol/μL)

Solutions of 17 amino acids in five concentrations are available from Agilent for calibration curves. Divide each 1-ml ampoule of standards p/n 5061-3330 through 5061-3334 into 100-μl portions in conical vial inserts. Cap and refrigerate aliquots at 4 °C.

Extended Amino Acid (EAA) Stock Solution:

Weigh 59.45 mg asparagines, 59.00 mg hydroxyproline, 65.77 mg glutamine, and 91.95 mg tryptophan into a 25-mL volumetric flask. Fill halfway with 0.1 N HCl and shake or sonicate until dissolved. Fill to mark with water for a total concentration of 18 n mol/μl of each amino acid. For high-sensitivity EAA stock solution, take 5 mL of this standard-sensitivity solution and dilute with 45 ml water (1. 8 n mol/μl). Solutions containing extended standards are unstable at room temperature. Keep them frozen and discard at first signs of reduced intensity.

Internal Standards (ISTD) Stock Solution

For primary amino acids, weigh 58.58 mg norvaline into a 50-mL volumetric flask. For secondary amino acids, weigh 44.54 mg sarcosine into same 50-mL flask. Fill halfway with 0.1 N HCl and shake or sonicate until dissolved, then fill to mark with water for a final concentration of 10n mol each amino acid/μl (standard sensitivity). For high-sensitivity ISTD stock solution, take 5 mL of standard sensitivity solution and dilute with 45 mL of water. Store at 4°C. Calibration curves may be made using two to five standards depending on experimental need. Typically 100pmol/μL, 250 pmol/μL, and 1 nmol/μl are used in a three-point calibration curve for "standard sensitivity" analysis. The following tables should be followed if internal standard or other amino acids (for example, the extended amino acids) are added. Table 2 describes "standard sensitivity" concentrations typically used in UV analysis.

Distinct Method Parameters

The LC Pump

- Compressibility (×10-6 bar) A: 35, B: 80
- Minimal Stroke A, B: 20 μL

The Automated Liquid Sampler (ALS): Online Derivatization

Depending on the auto sampler model, the automated online derivatization program differs slightly·

G1376C Well Plate Automatic Liquid Sampler (WPALS), with Injection Programme

1. Draw 2.5 μl from borate vial (Agilent p/n 5061-3339).
2. Draw 1.0 μl from sample vial.
3. Mix 3.5 μl in wash port 5 times.
4. Wait 0.2 min.
5. Draw 0.5 μl from OPA vial (Agilent p /n 5061-3335).

6. Mix 4.0 μl in wash port 10 times default speed.
7. Draw 0.4 μl from FMOC vial (Agilent p/ n 5061-3337).
8. Mix 4.4 μl in wash port 10 times default speed.
9. Draw 32 μl from injection diluents vial.
10. Mix 20 μl in wash port 8 times.
11. Inject.
12. Wait 0.1 min.
13. Valve bypass.

G1329A Automatic Liquid Sampler (ALS), with Injection Programme:

1. Draw 2.5 μl from borate vial (Agilent p /n 5061-3339).
2. Draw 1.0 μl from sample vial.
3. Mix 3.5 μl in air default speed 5 times.
4. Wait 0.2 min.
5. Draw 0.5 μL from OPA vial (Agilent p/n 5061-3335).
6. Mix 4.0 μL in air, 10 times default speed.
7. Draw 0.4 μL from FMOC vial (Agilent p/n 5061-3337).
8. Mix 4.4 μL, 10 times default speed.
9. Draw 32 μL from injection diluent vial
10. Mix 20 μL in air default speed 8 times.
11. Inject.
12. Wait 0.1 min.
13. Valve bypass

The location of the derivatization reagents and samples is upto the analyst, and ALS tray configuration. Using the G1367Cwith a 2 × 56 wellplate tray (G2258-44502), the locations were:

- Vial 1: Borate buffer
- Vial 2: OPA
- Vial 3: FMOC
- Vial 4: Injection Diluents
- P1-A-1: Sample

The Thermostatted Column Compartment (TCC)

Left and right temperatures are set at 40 °C. Enable analysis when temperature is within ± 0.8 °C. See Table 5 for whichheat sink to use.

The diode array detector (DAD)

Signal A

- 338 nm, 10 nm bandwidth, and reference wavelength390 nm, 20 nm bandwidth.

Signal B

- 262 nm, 16 nm bandwidth, and reference wavelength324 nm, 8 nm band widths.

Signal C

- 338 nm, 10 nm bandwidth, and reference wavelength390 nm, 20 nm band widths. Programmed to switch to262 nm, 16 nm bandwidth, reference wavelength 324 nm, 8 nm bandwidth, after lysine elutes and before hydroxy proline elute. Signal C is determined by examining signal A and B time frames between peaks 20 and 21, then choosing a suitable point in time to switch wavelengths. Once switch time is established and programmed into the method, signal A and Bare optional.

Peak Width Settings Were:

- >0.01 min for the Rapid Resolution High Throughput(RRHT) 1.8-μm column methods.
- >0.03 min for the Rapid Resolution (RR) 3.5-μm column methods.
- >0.03 min for the Traditional High Resolution 5-μm column methods.

Columns and Guard Cartridges

For Agilent ZORBAX Eclipse plus C18 columns and recommended guard cartridges. A Cartridge Hardware Kit (p/n 820888-901) is needed to house the guard cartridge. Cartridges should be flushed with a few milliliters of mobile phase B before connecting to column for equilibration.

Results and Discussion

The scalability of Agilent ZORBAX Eclipse plus C18 stationary phase columns between three particle sizes (5, 3.5, 1.8 μm) allows for many fine tuned methods. As summarized in Table 1, the analyst can choose a column or method based on analysis time, resolution factors, solvent usage, or instrument model. Each of the three categories has its own method parameters. Of particular importance is the mobile phase flow path configuration for the smaller volume columns. Truly scaling a gradient method between different column diameters and lengths is somewhat complicated, especially because different flow rates produce different gradient delay times, which is the time it takes the gradient to travel from its formation point to the column head. Besides flow rate, the delay time is determined by the flow path volume between the gradient formation point and the head of the column. The delay volume can vary from instrument to instrument. The specific flow paths used to produce all the chromatograms in this application note are listed in Table 5. Instead of trying to exactly scale the methods, a simpler approach was used with the flow path volumes listed in Table 5 and flow rates that produced a good separation. For example, the 4.6 × 50 mm, 1.8 um separation has a flow rate of 2 mL/min instead of 1. 5 mL /min, which is the equivalent flow rate, used for the longer 100, 150 and 250 mm, 4.6 mm id columns.

1. Prepare the HPLC system

a) Prepare Mobile PHase A and B

b) Set up the HPLC System
c) Connect to the mass spectrometer

2. Prepare the MS system

a) Perform a System Suitability Test
b) Review the test results
c) If necessary, update the acquisition and quantitation methods with the retention times

3. Perform the sample assay

a) Create a project folder
b) Load the autosampler
c) Perform the sample assay.

References

Allan G Gornall, Charles J Bardawill and Maxima M David (1948). Serum protein determination. 751 -766. Determination of serum proteins by means of the Biuret reaction.

Angelika Gratzfeld-Huesgen (1999). Sensitive and Reliable Amino Acid Analysis in Protein Hydrolysates using the Agilent 1100 Series HPLC Agilent Pub. # 5968-5658EN

Bates L S, Waldeen R P and Teare I D (1973) Plant Soil 39 205.

Bradford, M. M. (1976) *Anal. Biochem*. 72, 248–254.

Bradford M. M. (1976) Anal. Biochem. A rapid and sensitive method for the quantitation of microgram quantities of protein utilizing the principle of protein-dye binding. 72 248-254.

Burnette W N (1981) *Anal Biochem* 112 195.

Chinard E P (1952) *J boil Chem* 199 91.

Christine V. Sapan, Roger L, Lundblad. and Nicholas C. Price (1999). *Biotechnol. Appl. Biochem*.29, 99–108.

Cliff Woodward, John W Henderson Jr. and Todd Wielgos (2007) High-Speed Amino Acid Analysis (AAA) on Sub-Two Micron Reversed-phase (RP) Columns Agilent Pub. #5989-6297EN

Dierckx, S., Boeve, K., Van Camp, J. and Huyghebaert, A. 2006. Proteins, Peptides, and Amino Acids Analysis in Food. Encyclopedia of Analytical Chemistry

Hartree E.E. (1972). Anal. Biochem. 48:422 (This modification makes the assay linear over a larger range than the original assay)

Herbert Godel, Petra Seitz, and Martin Verhoef (1992)LC-GC International, 5(2), 44-49.

Horn M J, Jones D B and A E (1946) J Biol Chem 166 313.

http://biocyclopedia.com/index/plant_protocols/amino_acids_proteins/estimation_of_tryptophan.php

http://biocyclopedia.com/index/plant_protocols/amino_acids_proteins/silver_staining_of_proteins.php

http://www.biocyclopedia.com/index/plant_protocols/amino_acids_proteins/estimation_of_proline.php

http://www.biocyclopedia.com/index/plant_protocols/amino_acids_proteins/estimation_of_total_free_amino_acids.php

https://books.google.co.in/books?isbn=1441997857

https://books.google.co.in/books?isbn=8122409768

https://books.google.co.in/books?isbn=8122409768

https://books.google.co.in/books?isbn=8122409768

https://books.google.co.in/books?isbn=8122409768

https://books.google.co.in/books?isbn=8122409768

https://pdfs.semanticscholar.

https://www.gelifesciences.com/gehcls_images/GELS/Related%20Content/Files/1314716762536/litdoc80634334_20161012161457.pdf

https://www.lssc.edu/faculty/barbara_c_gutierrez/Documents%20%20Downloads/Exercise%2003%20Qualitative%20Analysis.pdf

https://www.scribd.com/document/59193908/Assay

Jason Greene, John W. Henderson Jr., John P. Wikswo, "Rapid and Precise Determination of Cellular Amino Acid Flux Rates Using HPLC with Automated Derivatization with Absorbance Detection" Agilent Pub.# 5990-3283EN (2009)

Jayaraman J. (1981) In: Laboratory Manual in Biochemistry. Wiley Eastern Limited New Delhi p 61-63.

John W. Henderson Jr, Robert D. Ricker, Brian A. Bidlingmeyer, and Cliff Woodward, (2000). "Rapid, Accurate,Sensitive and Reproducible HPLC Analysis of Amino Acids" Agilent Pub. # 5980-1193E

Lowry, 0. H., Rosebrough, N. J., Farr. A. L. and Randall. R. J. (1951) J. Biol. Chem.193, 265-275.

Lowry, O.H., Rosebrough, N.J., Farr, A.L., and Randall, R.J. (1951) J.Biol.Chem 193: 265 (The original method).

Merril C R, Danan M and Goldman D (1981) Anal Biochem110 201.

Mertz E T, Jambunathan R and Misra P S (1975) In: protein quality agricultural Research Stn Bull No. 7 Purdue Univ USA p 9.

Mertz E T, Jambunath R and Misra P S (1975) In: Protein Quality, Agric Experiment Stn Bull No 70 Purdue Uni USA p 11.

Misra P S, Mertz E T and Glover D V (1975) Cereal chem. 52 844.

Moore S and Stein W H (1948)inMethods in Enzymol (Eds:Colowick, S P and Kaplan N D) Academic Press New York 3 468.

Oakley B, Kirsch D and Morris N R (1980) Anal Biochem105, 361.

org/89c8/264e9e24b0b374d0cbbfb018e14b35ed2188.pdf

Poehling H M and Neuhoff V (1981) Electrophoresis2, 141.

Rainer Schuster(1988).J. Chromatogr, 431, 271-284.

Rainer Schuster and Alex Apfel, Hewlett-Packard(1986)App. Note, Pub. # 5954-6257.

Theymoli Balasubramanian and Sadasivam S (1987) Plant Foods Hum Nutr 37 41.

Theymoli Balasubramanian and Sadasivam S (1987) Plant Foods Hum Nutr 37 41.

Theymoli Balasubramanian and Sadasivam S (1987) Plant Foods Hun Nutr 37 41.

Towbin H, Staehelin T and Gordon J (1979) Proc Natl Acad Sci USA 76 4350.

Wilson, K. and Walker, J. (2000).Practical Biochemistry: Principles and Techniques", Cambridge University Press.

www.che.iitb.ac.in/courses/uglab/cl431/bl301-proteinassay.pdf

www.ibt.lt/uploads/file/bvtl-2/assays_3.pdf

www.ibt.lt/uploads/file/bvtl-2/assays_3.pdf

www.uobabylon.edu.iq/eprints/pubdoc_3_1536_218.doc

3

Chapter

Extraction and Quantification of Phenols

Phenolic compounds are important components of many fruits, vegetables, and beverages, in which they contribute to color and sensory properties such as bitterness and astringency. Epidemiological studies have shown that consumption of foods and beverages rich in Phenolic content is correlated with reduced incidences of heart disease.

One possible explanation is that Phenolic compounds slow the progression of atherosclerosis by acting as antioxidants toward low density lipo proteins (LDL). The oxidative modification of LDL is a critical step in the development of atherosclerosis, and prevention of this step is thought to slow the progression of the disease. Studies have shown that Phenolic compounds contained in fruits are potent inhibitors of the in vitro oxidation of LDL. There are some studies suggesting that Phenolic compounds have antioxidant activity in vivo. Therefore the primary objective of this investigation was to determine the level of total Phenolic, the concentration of each Phenolic compound, and the variability of Phenolic compounds in leaf and fruit.

3.1 Estimation of Total Phenols by Folin-Ciocalteu Colorimetric Methods

Phenols, the aromatic compounds with hydroxyl group are widespread in plant kingdom. They found in all parts of the plants. Grains containing high amount of polyphenols are resistant to disease and pest in plants. Grains containing high amount of compounds like tannins, flavonols etc. total Phenols estimation can be carried out with the Folin- Ciocalteau reagent.

Principle

Phenols react with Phosphomolybdic acid in Folin-Ciacalteau reagent in alkaline medium and produce blue colored complex (molybdemm blue).

Reagents

- Ethanol: 80 per cent in distilled water (w/v)
- Folin and Ciocalteu'sPhenol reagent: 1 N.
- Sodium carbonate solution: 20 per cent with distilled water.
- Standard Phenol solution: 100 mg of catechol is dissolved in 100ml of

distilled water. Form this solution 10 ml is taken and this is diluted to 100 ml with distilled water. The concentration of this solution will be 100 µg / ml (Working standard).

Procedure

1. Weigh exactly 0.5 to 1.0 gm of sample and grind with a pestle and mortar in 10 times volume of 80 % ethanol.
2. Centrifuge the homogenate at 10,000 rpm for 20 min. save the supernatant. Re extracts the residue with five times the volume of 80% ethanol, centrifuge and pool the supernatants.
3. Evaporate the supernatants to dryness.
4. Dissolve the residue in a known volume of distilled water (5 ml).
5. Pipette out different aliquots (0.2 to 2 ml) in to test tubes.
6. Make up the volume in each tube to 3 ml with water.
7. Add 0.5 ml of folin-Ciocalteau reagent
8. After 3 min, add 2 ml of 20% sodium carbonate to each tube.
9. Mix thoroughly. Place the tubes in boiling water bath exactly 1min, cool and measure the absorbance at 650nm against the reagent blank.
10. Prepare the standard curve using the different concentration of catechol.

Notes

1. If any white precipitate is observed on boiling, the color may be developed at room temperature for 60 min.
2. Express the result in terms of catechol or any other Phenol equivalent used as standard.

References

1. MalickC P and Singh M B (1980) In: *Plant enzymology and Histo Enzymology* Kalyani Publishers New Delhi p 286.
2. Singleton V L and Rossi J A Jr. (1965) *AMER. J. Enol. Viticult.* Colorimetric of total Phenolics with PhosphomlybdicPhosphotungstic acid reagent. **16** 144-58.
3. http://www.mdpi.com/1420-3049/18/6/6852/htm

3.2 Estimation of Tannins

Tannins and tannin like substances are widespread in nature and are probably present in all plant materials. Those are poly Phenolic compounds divided into two main groups – hydrolysable and condensed.

Hydrolysable tannin contains a polyhydric alcol usually, if not always, glucose esterified with gallic acid or with hexahydroxydiphenic acid. Condensed tannins are mostly flavonols are probably polymers of flavan 3-ol (catechin) and these cannot be hydrolysed to simple components.

The tannins are estimated by the following two methods:

a) **Folin- Denis Method**: This is based on the non- stoichiometric oxidation of the two molecules containing a Phenolic hydroxyl group.

b) **Vanillin Hydrochloric method** Vanillin method is specific for dihydroxyPhenols and is particularly sensitive for meta-substituted, di and tri hydroxybenzene containing molecules.

3.2.1 Folin- Denis Method

Principle

Tannin-like compounds reduce Phosphotungstomolybdic acid in alkaline solution to produce a highly colored blue solution, the intensity of which is proportional to the amount of tannins. The intensity is measured in a spectrophotometer at 700 nm.

Reagents

➢ **Folin- Denis reagents**

Dissolve 100 g sodium tungstate and 20 g Phosphomolybdic acid in 750 ml distilled water in a suitable flask and add 50 ml Phosphoric acid. Reflux the mixture for 2 h and make up to one liter of distilled water. Protect the reagent from exposure to light.

➢ **Sodium carbonate solution**

Dissolve 350 g of sodium carbonate in one liter of water at 70-80°C. Filter through glasswool after allowing it to stand overnight.

➢ **Standard tannic acid solution**

Dissolve 100 mg of tannic acid solution to 100 ml with distilled water.

➢ **Working standard solution**

Dilute 5 ml of the stock solution to 100 ml with distilled water. One ml contains 50 μg of tannic acid.

Procedure

1. **Extraction of tannic:**

   Weigh 0.5 g of the powdered material and transfer to a 250 ml conical flask.. Add 75 ml distilled water. Heat the flask gently and boil for 30 min. centrifuge at 2,000 rpm for 20 min and collect the supernatant in 100 ml volumetric flask and make up the volume.

2. Transfer 1 ml of the sample extract to a 100 ml volumetric flask containing 75 ml water.

3. Add 5 ml of Folin- Denis reagent, 10 ml of sodium carbonate solution and dilute to 100 ml with water.

4. Shake well, read the absorbance at 700 nm after 30 min.

5. If absorbance is greater than 0.7 make a 1 + 4 dilution of the sample.

6. Prepare the blank with water instead the sample.

Calculation

Calculate the tannin content of the samples as tannic acid equivalents from the standard graph.

3.2.2 Vanillin Hydrochloride Method

Principle

The vanillin reagent will react with any Phenol that has an unsaturated resorcinol or Phloroglucinol nucleus and forms a colored substituted product which is measured at 500 nm.

Reagents

- **Vanillin hydrochloride reagent**

  Mix equal volumes of 80% hydrochloric acid in methanol and 4% vanillin in methanol. The solution must be mixed just before use, and avoid using even if it is slightly colored.
- **Catechin stock standard solution**

  Prepared a standard solution containing 1 mg catechin / ml methanol
- **Working standard**

  Dilute the above stock solution ten times. (10 ml to 100 ml or 100µg /ml)
- **Preparation of the extract**

  Extract 1 gm of ground seed in 50 ml methanol. Mix occasionally by swirling. After 20 to 28 hr, centrifuge and collect the supernatant.

Procedure

1. Pipette out 1 ml of the supernatant
2. Quickly add 5 ml of vanillin hydrochloride reagent alone.
3. Prepare a standard graph with 20- 100 µg catechin using the diluted stock solution.
4. Prepare the blank with vanillin hydrochloric acid alone.
5. Prepare the standard graph with 20 to 100 µg catechin using the diluted stock solution.

Calculations

From the standard graph, calculate the amount of catechin, i.e. tannin in the sample as per the absorbance values and express the result as catechin equivalents.

3.3 Estimation of Lignins

Lignins are Phenolic polymers present in the cell walls of plants which are responsible together with cellulose, for the stiffness and rigidity of the plant stems. Lignins areespecially associated with woody plants, since up to 30 % of the organic matter of trees consists of lignin. Lignin acts as Physical barrier against invading pathogens.

Principle

Refluxing the sample material with acid detergent solution removes the water-solubleand material other than the fibrous component. The left- out material is weighed after filtration, dried, treated with 72 % H_2SO_4 and filtered, dried and shed. The loss of water on ignition gives the acid detergent lignin.

Reagents

- Acid detergent solution: Dissolve 20n g of cetyl-trimethyl ammonium bromide in one liter of 1 N sulphuricaicd.
- 72% H_2SO_4
- Acetone
- Round bottom flask and refluxing set
- Muffle furnace
- Sintered glass crucible

Procedure

A. Acid Detergent Fiber (ADF)

1. Place 1 g of powdered sample in around bottom flask and 100 ml of acid detergent solution. Heat to boil in 5 to 10 min. reduce heat to avoid foaming as boiling begins. Reflux for 1 hr after the onset of boiling. Adjust boiling to slow, even level.
2. Remove container, swirl and filter the contents through a pre-weighed sintered glass crucible (G-2) by suction and wash with hot water twice.
3. Wash with acetone and break up the lumps. Repeat acetone washing until the filter is colorless
4. Dry at 100°C for overnight.
5. Weigh after cooling in desiccators.
6. Express ADF content in percentage. W/S X 100, where W is the weight of the fiber and S is the weight of sample.

B. Determination of Acid Detergent Lignin (ADL)

1. Transferto a 100 ml beaker with 25- 50 ml sulphuric acid. Add 1 g asbestos. Allow it to stand for 3 hr with intermittent stirring with a glass rod.
2. Dilute the acid with distilled water and filter with pre-weighrd Whatman No. 1 filter paper. Wash the glass rod and the residue several times to get rid of the acid.
3. Transfer the filter paper to a pre-weighed silica crucible and ash the filter paper with the content in a muffle furnance at 550°C for about 3 hr.
4. Cool the crucible in a desuccator and weigh. Calculate the ash content.
5. For blank take 1 g asbestos, add 72 % H_2SO_4 and follow the step 2 to 5.

Calculation

$$\% \text{ (ADL)} = \frac{\text{Weight of 72 \% } H_2SO_4 \text{ Washed fiber (Test – Asbestos blank)} - \text{Ash (Test – Asbestos blank)}}{\text{Weight of sample}} \times 100$$

Notes

1. NADF = Hemicellulose + Cellulose + Lignin + Minerals

   ADF = Cellulose + Lignin + Minerals

   Hemicellulose = NDF - ADF

   Cellulose = ADF – Residue after extraction with 72 % H_2SO_4

   Lignin = Residue after extraction with 72 % H_2SO_4 – Ash
2. Wet materials can also be used for lignin estimation, provided the wet sample is equivalent to 1 g of dry materials.
3. If the weight loss of asbestos blank on ashing is below 0.002 g/g, discontinue the determination of blank.

3.4 Estimation of Capsaicin

Capsaicin is a protoalkaloid which is responsible for the pungency of chilies. The quality of chili fruits, extracts or oleoresins is determined by the capsaicin content.

3.4.1. Estimation of Capsaicin by Colorimetric Method

Principle

The Phenolic group in capsaicin reduces Phosphomolybdic acid to lower acids of molybdenum. The resulting component is blue in color is read at 650nm. The color intensity is directly proportional to the concentration of capsaicin.

Materials

- 0.4% Sodium Hydroxide
- 3% Phosphomolybdic acid
- Dry Acetone (add about 25g anhydrous sodium sulphate to 500ml acetone of analytical grade at least one day before use)
- Stock Standard Capsaicin Solution

Dissolve exactly 50 mg capsaicin in 50 ml of 0.4 sodium hydroxide solution (1, 000µg/ml).

- Working standard:

Dilute 10ml of the stock standard to 50 ml with 0.4% sodium hydroxide solution (200µg/ml)

Procedure

1. Weigh 0.5 g dry chili powder into a glass-stoppered test tube or volumetric flask.

2. Pipette out 10 ml dry acetone into the flask and shake well it for 3h in a mechanical shaker. Let the contents settle down or centrifuge at 10,000 rpm for 10 min.
3. Pipette out 1 ml of the clear supernatant into a test tube and evaporate to dryness in a hot water bath.
4. Dissolve the residue in 5ml of 0.4% sodium hydroxide solution.
5. Add 3ml of 3% Phosphomolybdic acid.
6. Shake the contents and let stand for 1hr.
7. Filter the solution quickly into centrifuge tubes to remove any floating debris.
8. Centrifuge at about 5,000 rpm for 10- 15 min.
9. Transfer the clear blue colored solution directly into the cuvette and read the absorbance at 650nm.
10. Run the reagent blank along with the test sample
11. Prepare a standard graph using 0 – 200 µg capsaicin simultaneously i.e., pipette out 0.2, 0.4, 0.6, 0.8 and 1 ml of working standard solution and proceed as above.

3.4.2. Estimation of Capsaicin by Spectrophotometric Method

Principle

Capsaicin is extracted from red pepper (Chilies) with ethyl acetate and made to react with ethyl acetate solution of vanadium oxycholride. Then it is read at 720nm. This method is sensitive and is useful to measure small quantities (less than 0.05%).

Materials

- Vanadium oxychoride (0.05%) in ethyl acetate
- Pure capsaicin (0.01%) in ethyl acetate (10 mg in 100ml)

Procedure

1. Grind the sample well to pass through No.40 sieve.
2. Place 2g sample in 100ml volumetric flask.
3. Let it stand for 24 hr to extract (otherwise reflux the contents for 2.5 hr) and then make up to volume.
4. Dilute 1ml of extract to 5ml with ethyl acetate.
5. Add 0.5ml vanadium oxychloride solution (just before reading) and shake well.
6. Read the absorbance at 720nm in a spectrophotometer.
7. Substract the absorbance value.
8. Prepare a standard curve using 0.5, 1.0, 1.5, 2.0 and 2.5 ml of standard capsaicin solution containing 50, 100, 150, 200 and 250 µg capsaicin respectively.

Calculation

$$\% \text{ capsaicin} = \frac{\mu g \text{ capsaicin}}{1000 \times 1000} \times \frac{100}{1} \times \frac{100}{2}$$

$$= \frac{\mu g \text{ Capsaicin}}{200}$$

3.5. Estimation of Chlorogenic Acid

Phenolic compounds are widely distributed in plants. They contribute to the color and flavor of plant parts. The presence of Phenolic compounds in the oilseeds and grains pose some nutritional problems. Among the Phenolic compounds, chlorogenic acid, found in the range of 2g/100g of defatted sunflower meal is a matter of immediate concern. Breeding for low cholorogenic acid from the defatted meal are, today, the major areas of sunflower research. A rapid method to estimate chlorogenic method is described.

Principle

A Chlorogenic acid is extracted with alcohol dried and dissolve in acetone. It is reacted with titration ion to form a colored complex which is measured at 450nm.

Materials

- Titanium reagent: 20% TiCl4 in conc. HCl
- Standard 25 – 200 μg/ml Chlorogenic Acid in acetone
- Acetone
- 80% Ethanol
- 2.5 N HCl

Procedure

1. Reflux twice a known quantity of defatted sunflower meal in 80% ethanol (adjusted *pH* 4.0 with 2.5 N HCl) for 30 min (125ml to 1g meal).
2. Discard the precipitate and collect 250ml the extract.
3. Remove 0.5 ml sample and dry in a vacuum oven at 50°C and 700 nm pressure for 2 hr.
4. Dissolve dried extract in 4.75ml of acetone
5. Add 0.25ml of $TiCl_4$
6. Read the color at 450 nm against a reagent blank (acetone plus $TiCl_4$)
7. Similar treat the standards with $TiCl_4$ and read the color intensity.
8. Draw a standard curve and find out the chlorogenic content in the sample.

Calculations

Express chlorogenic acid content as g per 100g sample.

Notes

1. If the titanium chloride is not available, this estimation can be perforated with Folin-Ciocalteau reagent. Dissolve the dried extract in 5 ml ethanol

and proceed for Folin-Ciocalteau method (refer Phenol estimation). Similarly, prepare the standard in 80% ethanol and develop the color with Folin-Ciocalteau reagent. In this method, there is a possibility of interference of tyrosine and Phenyl amine.

2. The titanium chloride method can be followed for the other Phenols. Too. But the wavelength of maximum absorbance differs. The total Phenols can be measured at 410 nm, catechol and catechin at 430nm.

References

file:///C:/Users/temp1001/*Downloads/1194-1967-1-P*B.pdf

Goering H D and Van Soest P J (1975) Forage Fiber Analysis U S Dept of Agriculture Agricultural Research Service Washington DC.

http://pubs.acs.org/doi/abs/10.1021/ac60019a007

http://vc-dei.vlabs.ac.in/?page_id=1036

http://www.ajevonline.org/content/16/3/144.short

http://www.mdpi.com/1420-3049/18/6/6852/htm

https://s3.amazonaws.com/academia.edu.documents/34289991/Determination-Structural-Carbohydrates-

https://www.cabdirect.org/cabdirect/abstract/19641407080

https://www.ncbi.nlm.nih.gov/pubmed/15467248

Michael Eskin, M A Hochn E and Frenkel C (1978) J Agric Food Chem

Official Methods of Analysis of the AOAC 12thEdn (1975) p 138.

Palacio J J R (1977) JAOAC 60 970.

Quagliotti L (1971) Hortic Res 11 93.

Robert E B (1971) Agro J 63 511.

Schander S H (1970) In: Methods in Food Analysis Academic Press New York p 709

TheymoliBalasubramanian, Raj D, Kasthuri R and Rengaswami P (1982) Indian J Hortc39 239.

4

Chapter

Extraction and Quantifications Procedures of Lipids

Introduction

Biological lipids are a chemically diverse group of compounds, the common and defining feature of which is their insolubility in water. Fats and oils are the principal stored forms of energy in many organisms. Phospholipids and sterols are major structural elements of biological membranes. Other lipids, although present in relatively small quantities, play crucial roles as enzyme cofactors, electron carriers, light absorbing pigments, hydrophobic anchors for proteins.

The present chapter aim to carry out for the estimation of (lipids) sterols in edible fats and oils. Samples of vanspati ghee which were most commonly available in the markets were subjected to sterol estimation. The sample included Dalda, Tullo, Habib, Fauji, Kisan, ACP, Latif, Family, Zaiqa and Handi, Vanaspathi. Dalda Tullo and Habib vanaspati contended relatively less amount of cholesterols (10, 14 and 15 mg/g respectively) while, fauji, Kisan, ACP and Latif Vanaspathi had the intermediate values of cholesterol (24, 25, 29 and 30 mg/g respectively). Family, Zaiqa and Haldi Vanaspati contended maximum amount of chloresterol (33, 37 and 40 mg/g respectively). Dalda the vanaspathi had the least (10mg/g) while, Handi Vanaspathi had the highest (40 mg/g) amount of cholesterol. Among the others samples butter and Haleeb desi ghee contents the high amount cholesterol (27, 7 mg/g respectively).While, milk fat, beef tallow and fish fat were low in cholesterol contents (3, 5.1 and 4 mg/g respectively). Beer fat had the immense quantity of cholesterol (80 mg/g). The sterol content of edible oils i.e. (23 mg/g), soyabeen (9mg/g) repseed (5mg.g) and coconut (0.8mg/g) were reported less as comapired to the steriol contents of mustered oil (64mg/g).

4.1 Estimations of Sterols

As the plant sterols lower down the level of cholesterol in blood thus, the level of phytosterol was also estimated in mustard oil, coconut oil, corn oil, rapseed (Canola) oil and soyabeen.

Principle

The Liberman-Burchard reagent reacts with sterols to produce a characteristics green color whose absorbance is determined on spectrophotometer at 640nm.

Materials

- Sample collection and preparation:
  - Samples of edible fats and oils were collected from local markets, weigh the sample accurate 1 gm. Chloroform used as a solvent to make the volume 10 ml. dissolve the sample completely by starring and dilute to 10 times. Take 3 ml from the diluted sample solution and measure the absorbance after adding the reagent at 640nm.
- Preparation of standard solution:
  - 10 mg of standard cholesterol dissolve in 10ml of chloroform, shake well.
- Liberman –Burchard Reagent:
  - 0.5 ml of sulfuric acid dissolves in 10 ml of acetic anhydride. Covered and kept in ice bucket.

Procedure

1. Pipette out the standard cholesterol solution in different test tube and mark them appropriately.
2. Add standard cholesterol solution as 0.5, 1, 1.5 and 2.
3. Then add 2 ml of Liberman- Burchard reagent in each tubes.
4. Make the final volume of adding chloroform
5. Cover the all tubes by covering with black carbon paper and kept in dark for 15 minutes.
6. Set the zero of spectrophotometer with blank at 640 nm.
7. The absorbance of all standards were determined on spectrophotometer and standard graph was plot.
8. Finally, 3ml of sample solutions were taken and their absorbances were determined on spectrophotometer after adding the Liberman-Burchard reagent and chloroform.

Results

The calibrations curves we have obtained using the reagent and the method describe above obey satisfactorily the law of Lambert and Beer.

4.2 Estimation of Saponification Value

Introduction

Saponification value indicates the average molecular weight of a fat or oil. The saponification value may be defined as the number of milligrams of caustic potash required to neutralize the fatty acids obtained by complete hydrolysis of one gram of oil or fat. Thus saponification value gives us information whether an oil or fat contains high proportion of lower or higher fatty acids. For e.g., butter has a large proportion of lower fatty acids than lard and tallow, and has high saponification value. Coconut oil also has a comparatively higher saponification value.

Applications of saponification value: Saponification value gives us an idea about the molecular weight of fat or oil.

Materials

- Hydrochloric acid 0.5 N, accurately standardized.
- Alcoholic KOH:
  - Dissolve the 40g of KOH in 1 liter of distilled alcohol keeping the temperature below 15°C while the alkaline is being dissolved. This solution should remain clear.
- Phenolphthalein indicator 1 % in 95 % ethanol
- Air Condenser.

Procedure

1. Weigh 1 gm of oil and transfer into the round bottom flask
2. Fill the burette with 50ml of KOH solution and allow to stay for few minutes.
3. Prepare the blank also by adding alcoholic KOH and stay for same time as above.
4. Connect the air condenser to the flasks and boil them gently for an hour.
5. After the cooling the flask and condenser rinse down the inside of the condenser with a little distilled water and then remove the condenser.
6. Add about 1ml of indicator and titrate against 0.5 N HCl until the pink color is disappears.

Calculations

Saponifiaction value:

$$= \frac{28.5 \times (Titrate\ value\ of\ blank - Titrate\ value\ of\ sample}{Weight\ of\ sample(g)}$$

Notes

1. Alcohol is inflammable, use electrical heating
2. Alcohol should not get dried up during saponification. Effective cooling vapour is essential
3. Clarity and homogeneity of the test solution of complete saponification.

4.3 Estimation of Free Fatty Acids

A small quantity of free fatty acids is usually present in oils along with the triglycerides. The free fatty acid content is known as acid number/ acid value. It increases during storage. The keeping quality of oil therefore relies upon the free fatty acid content.

Principle

The free fatty acid in oil is estimated by titrating it against KOH in the presence of phenolphthalein indicator. The acid number is defined as the mg of KOH required to neutralize the free fatty acids present in one gram of sample. However, the free fatty acid content is expressed as oleic acid equivalent.

Materials

- 1 % phenolphthalein in 95% ethanol
- 0.1 N Potassium hydroxide
- Neutral Solvent: Mix 25 ml ether, 25 ml 95% alcohol and 1ml of 1% phenolphthalein solution and neutralize with N/10 alkali.

Procedure

1. Dissolve 1 to 10 g of oil or melted fat in 50 ml of the natural solvent in a 250ml conical flask.
2. Add few drops of phenolphthalein
3. Titrate the contents against 0.1N potassium hydroxide
4. Shake constantly until a pink color which persists for fifteen seconds is obtained.

Calculations

$$\text{Acid value}\ (\text{mg KOH/g}) = \frac{\text{Titre value x Normality of KOH} \times 56.1}{\text{Weight of the sample}\ (\text{g})}$$

The free fatty acid is calculated as oleic acid using the equation 1ml N/10 KOH = 0.028g oleic acid.

Note

To find out the extract strength of KOH, prepare 0.1 N oxallic acid solution (630mg in 100ml water) and titrate against KOH wit phenolphthalein as indicator. Calculate the strength of KOH by the formula $V_1N_1 = V_2N_2$.

4.4 Determination of Iodine Value

Method for the determination of iodine value (g I2/100g) of edible oils and fats according to Wijs as indicated by AOAC Official Methods of Analysis (1984).

Principle

The oil contains both saturated and unsaturated fatty acids. Iodine gets incorporated into the fatty acid chain wherever the double bond of exist. Hence, the measure of idine absorbed by oil, gives the degree of unsaturation. Iodine value/ number is defined as the 'g' of iodine absorbed by 100g of the oil.

Materials

- Sample:
  - Fats and oils, approx 25g/expected iodine value
- Substance
  - C-C double bonds, -CH=CH-
- Chemicals

- 15 ml carbon tetrachloride, CCL_4; 25 ml Wijs solution; 100ml of water, 15 ml 10% potassium iodide, KI
- Sodium thiosulphate, Na2 S2O3 C ($Na_2S_2O_3$) = 0.1mol/l.
- Wijs Solution: Dissolve 16.2 g iodine mono chloride (ICI)in a 1 liter volumetric flask with glacial acetic acid. store the Wijs solution in amber bottle sealed with paraffin until it is used. (Wijs solution is sensitive to temperature, moisture and light.)

Procedure

1. Dissolve the sample in CCl_4, add 25 ml of Wijs Solution and keep in the dark for approx one hour to complete the reaction.
2. Add deionized water and titrate excess iodine with sodium thiosulphate
3. Add few drops of starch as indicator and titrate until the blue color completely disappears.
4. Towards the end of titration, stopper the flask and shake vigorously so that any iodine remaining in solution in $CHCl_3$ is taken up by potassium iodide solution.
5. Run the blank without the sample.

Calculation

The quantity of thiosulphate required for blank minus the quantity required for sample gives thiosulphate equivalent of iodine absorbed by the fat or oil taken for analysis.

$$\text{Iodine number} = \frac{(B - S) \times N \times 12.69}{\text{Sample (g)}}$$

4.5. Estimation of Volatile Acid

Low-molecular mass carboxylic acids, (C_2-C_7mono carboxylic aliphatic acids) are important intermediates and metabolites in biological processes. These carboxylic acids are known as volatile fatty acids (VFAs) or short-chain fatty acids (SCFAs). Volatile fatty acid is present in environmental water in the range of 1 to 5000ppm and different methods have been reported for their determination. The methods involved distillation method, potentiometric titration, spectrophotometric and gas chromatographic method. Every methods had its advantage and disadvantages. The spectrophotometric method was sensitive for hydrogen carbonate alkalinity and phosphate ions, measured concentration of acetic acid were lower than nominal. The precision and accuracy of the spectrophotometric method were in the ranges 1.3 -14% and 82.1 to 104%.

Estimation of Volatile Acid by Titration Method

Principle

Straight distillation or steam distillation after acidifying the sample with 1:1 sulfuric acid results in the evaporation of volatile acids which on condensation yield liquefied volatile acids. These acids could be titrating against 0.1N alkali with phenolphthalein indicator to find the amount of volatile fatty acid.

Materials

- Dilute sulfuric acid 1: 1 mix one volume of Conc H_2SO_4 with one volume of water.
- Ferric chloride solution: Dissolve 82.5g FeC_{13} $6H_2O$ in one litre distilled water.
- Diatomaceous-Silica Filter Aid
- Magnesium Sulphate ($MgSO_4$. $7H_2O$)
- Standard Sodium Hydroxide, titrate 0.1 N: Dissolve 4 g of NaOH in 1 litre of distilled water and standard against standard acid.
- Phenolphthalein Indicator: 0.1% in 95% ethanol.

Procedure

1. Adjust the sample containing volatile fatty acid (Sludge and water) 200 to 1000ml to Ph3.5 with (1+1)sulfuric acid.
2. Add 6 ml of $FeCl_3$ solution/L sample equivalent to 500mg/l
3. Add 50g filter aid/l and mix well.
4. Filter by suction, using a Buchner funnel containing a filter paper freshly coated with a thin layer of filter-aid
5. Wash the residue thoroughly three or four times with water, and adjust the filter to pH 11 with NaOH solution.
6. Concentrate by evaporation to 150 ml and cool in a refrigerator.
7. Adjust the cool filtrate to ph4 with dilute H_2SO_4 and add it quantitatively and quickly to the distilled flask.
8. Add magnesium sulphate to a slight excess of saturation
9. Apply heat with a small flame to the flask until rapid evalution of volatile acids commences. This will prevent excessive increase in the volume of the mixture.
10. Steam distilled slowly so that about 200 ml of distillate is collected in 25min.
11. Increase the rate of distillation and continue until a total of 600 ml is collected.
12. Titrate the distilled against 0.1 N NaOH using phenolphthalein indicator.

Calculations

$$Mg/L\ volatile\ acids\ as\ acetic\ acid = \frac{M_l 0.1\ N\ NaOH \times 6000}{Sample\ (ml)}$$

Notes

The steam distillation method is more tedious and quite poor absolute recovery (53-58%) from the given matrix and rather elevated limit of quantification (LOQ) at 110 mg/L.

b. Determination of Acetic Acid by a Spectrophotometric Method

Introduction

In literature different methods for the determination of VFAs in environmental samples are reported. Traditionally, the VFA content in wastewaters and foods has been analyzed by titrimetric or gas chromatographic methods preceded by solvent extraction. The rapid colorimetric determination of organic acids is also in use.

Principle

The spectrophotometric procedure is based on the well known colorimetric ferric hydroxamate method for determination of carboxylic esters known as the Montgomery method

Materials

- Acetic acid
- ethylene glycol
- Sulfuric acid
- hydroxylamine hydrochloride
- 4.5 N NaOH
- Ferric chloride.

Procedure

1. Prepare standard solution of acetic acid 1200mg/l were diluted with distilled water to obtain appropriate concentrations in the range of 10 to 1200mg/l
2. In this study the procedure was modified as follows: aqueous sample (0.5 mL) was taken into a dry test-tube.
3. Add 1.5 ml of ethylene glycol reagent and 0.2 ml of 19.5 N sulfuric acid.
4. Heat the mixture about 3 minutes in a boiling water bath.
5. Cool the content in the test tube immediately by cooling water.
6. Add 0.5 ml of 10% hydroxylamine hydrochloride solution, 2ml of 4.5 N NaOH solution and 10 ml of 10 % ferric chloride solution.
7. Read the absorbance at 495nm.
8. Plot the graph against standard.

Calculations

The calibrations curves we have obtained using the reagent and the method describe above obey satisfactorily the law of Lambert and Beer.

Results

The method is based on estrification of the carboxylic acids present in sample and subsequent determination of the esters by the ferric hydroxamate reaction. All volatile fatty acids present in the sample are reported as their equivalent mg/Lof acetic acid.

4.6 Estimation of Oils From Oil Seeds

Principle

Fats are the tri-esters of glycerol, when it is in liquid are called oil. Seeds like groundnut, castor, sunflower, soyabeen, peanut etc cantinas the oils as reserve food materials for the embryo.

Materials

- Petroleum Ether
- Whatman No.2 filter paper
- Absorbant Cotton
- Soxhlet Apparatus

Procedure

Preparation of Sample

a) **Peanut:** Take 50 g of kernels in a drying dish and dry it at 130°C for not more than 20 min in a forced draft oven. Cool to room temperature and then pass through the nut slicer. Care is to be taken to prevent expressing of any oil while slicing. Mix the sliced sample well. Weigh accurately 2g into the filter paper fold.

b) **Sunflower sample**: Weigh about 50g of sunflower seeds into a large beaker. Add an equal weight of hyflow supercel or equivalent (dried at 130°C overnight before use). Mix well. Grind the mixture immediately.

c) **Soyabeen**: Dry the sample well either by keeping overnight in a warm room or in an oven so that the moisture is within 6-8%. Grind in a micro-sample mill with minimum exposure to air. Mix thoroughly weigh exactly 2g of the sample and place in the filter paper fold.

1. Fold a piece of filter paper in such a way to hold the seed meal. Wrap around a second filter paper which is left open at the top like a thimble. A piece of cotton wool is placed at the top to evenly distribute the solvent as it drops on the sample during extraction.
2. Place the sample packet in the butt tubes of the Soxhlet extraction apparatus.
3. Extract with petroleum ether (150 drops/ min) for 6 hours without interruption (for castor beans use hexan) by gentle heateing.
4. Allow to cool and dismantle the extraction flask. Evaporate the ether on a steam or water bath until no odour of ether remains. Cool at room temperature.

5. Carefully remove the dirt or moisture outside the flask and weigh the flask. Repeat heating until constant weight is recorded.

Calculations

$$\text{Oil in ground sample} = \frac{\text{Weight of oil (g)}}{\text{Weight of sample (g)}} \times 100$$

$$\text{Oil to dry weight basis} = \frac{\text{\% oil in ground sample}}{\text{100\% moisture in whole seed}}$$

Calculation for sunflower seeds:

$$\% = \frac{\text{Weight of oil (g)} \times 200}{\text{Weight of sample}}$$

References

AOAC Official Method of Analysis, 1984, Chapter 28.02.

Corrol K. K, P. M. Giovannetti M. W. Huff and D. C. Roberts (1984). Am. J. Clin. Nutr., 31: 131-21. Hypercholesterolemic effect of substituting soybean protein for animal protein in the diet of healthy young women.

Cox H E and Pearson D (1962) The Chemical Analysis of foods Chemical Publishing Co Inc New York p 420.

http://journals.plos.org/plosone/article?id=10.1371/journal.pone.0089643

http://pubs.acs.org/doi/abs/10.1021/ac60209a029

http://www.abacuslab.ru/media/115222/an%20325%20extraction%20of%20oils%20from%20oilseeds%20by%20accelerated%20solvent%20extraction%20(ase).pdf

http://www.jbc.org/content/164/2/725.full.pdf

http://www.sciencedirect.com/science/article/pii/0026265X83900309

https://tools.thermofisher.com/content/sfs/brochures/AN-325-extraction-of-oils-from-oilseeds-by-accelerated-solvent-extraction.pdf

https://www.ncbi.nlm.nih.gov/pmc/articles/PMC3933640/

Khan M. R. (1988) Text book of biochemistry. Carvan Book House Lathore.

Mehta S. L., Lodha and Sane P.V. (1983). Resent advances in plant Biochemistry, Indian Council of Agriculture Research New Delhi, 14-20.

Montgomeryh A C, Dymock J. F. Thom N. S. (1962). The Analyst. 87, 949. The rapid colorimetric determination of organic AIDS and their salts in sewage-sludge liquor.

Olmslead W H, Whitaker W M and Duden C W (1930) J Biol Chem 85 109.

Rand M C, Greenberg A E and Taras M J (eds) (1976) In: standard methods for the Examination o f waste water (14th Ed) American Public Health Association Washington p 529.

Sadasivam S., Manickam A (2005). Biochemical methods (2ed) p 22-23. New Age International (P) Ltd., Publishers. New Delhi.

Siedlecka E. M., Kumirska J., Ossowski T., Glamowski P., Gołębiowski M., Gajdus J., Kaczyński Z., Stepnowski P. (2008). Polish J. of Environ. Stud. Vol. 17, No. 3,351-356. Determination of Volatile Fatty Acids in Environmental Aqueous Samples.

William Horowiz (ed) (1975) official methods of Analysis of AOAC Association of official Analytical Chemists Washington 12th Ed p 490.

5

Chapter

Extraction and Quantification Procedures of Enzymes

Enzymes are proteins which catalyze a variety of reaction in the biological systems. The multifarious enzymes present in a living cell can be classified into six major groups based on the type of reaction they are catalyzed. These are:

1. **Oxidoreductase**: Transfer of electrons (hydride ions or H atoms)
2. **Transferases**: Group transfer reactions
3. **Hydrolases**: Hydrolysis reactions (transfer of functional groups to water)
4. **Lyases**:Addition of groups to double bonds, or formation of double bonds by removal of groups
5. **Isomerases**: Transfer of groups within molecules to yield isomeric forms
6. **Ligases**:Formation of C – C, C - O, C – S, C – N bonds by condensation reactions coupled to ATP cleavage.

Different techniques have been used to measure enzyme activities. There is no general method equally applicable to all enzymes. Many precautions have to be taken while handling enzymes, since several parameters like temperature, pH, salt concentration, presences of ions, etc. affect the enzyme activity. The optimal conditions of each enzyme have to be carefully worked out by the trial and error methods.

Here in this chapter, a few experiments are described by which simple to handle enzyme can be characterized and assayed. The procedure for measuring the activity of an enzyme is termed as assay. Fallowing some rooting methods are described on the basis if spectrophotometry.

5.1 Malate Dehydraogenase

(L walate: NAD+oxidoreductase EC 1.1.1.37)

Introduction

Malate dehydrogenase is one of the enzymes involved in TCA cycle. It catalyses the reversible conversion of oxaloacetic acid to malic acid.

$$\text{Oxaloacetic acid} + \text{NADH} \xrightleftharpoons{Mg^{2+}} \text{Malic acid} + \text{NAD}^+$$

This enzyme is also involved in carbon dioxide assimilation in C_4 plants. It is coupled with Pepcase in CO_2 assimilation in the chloroplast of the mesophyll cells. In this procedure NADPH is used as the coenzyme.

Principle

Malate dehydrogenase was estimated by the method of Gnanam and Francis (1976). Since it is an oxidoreductase involving nicotinamide adenine dinucleotide the decrease in absorbance is due to the oxidation of NADH is followed.

Materials

- Oxaloacetic acid (5µmole) 66 mg/50ml in distilled water.
- Magnesium chloride (10µmole) 203.5 mg/50 ml of distilled water.
- Tris-HCl Buffer (0.1M) pH 7.8
- NADH (0.4µmole) 5.32mg/10 ml of distilled water.

Procedure

1. Oxaloacetic acid, magnesium chloride, Tris HCl buffer, enzyme extract, all these are taken in a test tube each solution was contain in 0.5 ml and mixed well.
2. The spectrophotometer was settled to a zero absorbance at 340 nm without added NADH in the test against the blank in the reference cuvette.
3. Added NADH was quickly and mixed well.
4. Recorded the initial absorbance and absorbance every 30 seconds for at least 5 minutes.

Calculations

Calculate the enzyme activity as follows with decreased in absorbance for one minute.

µmoles NADH oxidized per minute 0.2ml enzyme extract

= absorbance decrease/min x 0.1613 x 3 (volume of the reaction mixture in ml)

Determine the protein content by the method of Lowery *et al.*in the enzyme extract. Compute the value for mg protein to calculate the specific activity.

5.2 Glutamate Dehydrogenase

Introduction

Glutamate dehydrogenase (GDH) occurs in almost all living organisms. In higher plants, GDH activity has been found in most species tested. The existence of the two distinct GDH enzymes in higher plants is now well documented. One is mitochondrial enzyme NAD, second is chloroplast NADP. The reaction catalyzed by this enzyme is given below.

$$\text{L- Glutamate} + H_2O + NAD^+(P) \rightleftarrows 2\text{– Oxoglutarate} + NH_4^+ + NAD(P)H + H$$

Principle

Glutamate dehydrogenase was estimated by the method of Delma Dohertry (1990). GDH is like other dehydrogenase assayed by following the oxidation of the reduced coenzyme, NADH or NADPH. These reduced coenzymes absorb light at 340 nm, which are most biological systems is uniquently uncluttered with interfering absorption by other compounds. Thus even in crude extracts the absorption of NADH at 340 nm is easily detected.

Materials

1. Potassium phosphate buffers 1.0 M.

- Stock solution: -
  a) 0.2 M solution of monobasic sodium phosphate (27.8 g in 1000 ml)
  b) 0.2 M solution of dibasic sodium phosphate (53.65 g of (NH_2PO_4.7H_2O or 71.7 g of Na_2HPO_4.12 H_2O in 100ml)

  8.5 ml of A, 91.5 ml of B was diluted to a total of 200 ml of distilled water.
- 2-Oxoglutarate 0.1 M:-

  14.6 g 2-oxo-glutarate was dissolved in one liter of distilled water.
- NH_4Cl (6.0 M): -

  53.5 g of NH_4Cl was dissolved in one litre of distilled water.
- NADH:-

  10 mg of NADPH was dissolved in 10 ml of water.
- NADPH: -

  10 mg of NADPH was dissolved in 10 ml of water.

Enzyme Extract

Extract 1gm of the plant material with 5ml of 100mm phosphate buffer Ph 7.5 containing 1mm disodium EDTA, 1mm dithioerythritol and 1% polyvinyl pyrrolidone (PVP) and centrifuge at 10,000 rpm for 30 minutes at 4°C. Collect the supernatant and use it for enzyme assay.

Procedure

1. 1ml of potassium phosphate buffer 7.0 and 7.8 are taken into the test tube.
2. Added 0.31 of 2-oxoglutarate, 0.5 ml of NH_4Cl and 0.12 ml of NADH solution.
3. The 0.2-ml of enzyme extract and 8 ml of water was added and mixed well.
4. Added 3 ml of water in black instead of 2- oxoglutarate.
5. This solution was incubated at 37°C for 15-30 minutes.
6. The values are recorded the changes in absorbance at 340 nm.

Calculation

The amount of NADH or NADPH oxidized was calculated from the molar extinction coefficient. Activities are expressed as n mole NAD (P) H oxidized per minute per mg protein.

GDH nmole of NAD (P) H oxidized/min/mg protein =

A_{340} X volume of assay solution X 1000 X time of incubation (min) X mg protein in enzyme extract used

5.3 Nitrate Reductase

The assimilatory reduction of nitrate by plants is a fundamental biological process in which a highly oxidized form of inorganic nitrogen is reduced to nitrite and then to ammonia.

$$NO_3^- + AH_2 \rightarrow NO_2^- + A + H_2O$$

The nitrate reducing system consists of nitrate reductase and nitrite reductase which catalyze stepwise reduction of nitrate to nitrite and then to ammonia. According to the specificity of electron donor two major types of nitrate reductase occur:

a) Ferredoxin-dependent nitrate reductase (blue-green algae)

b) Pyridine nucleotide-dependent nitrate reductase (higher plants)

Principle

Nitrate reductase (NR) is capable of utilizing the reduced form of pyridine nucleotides, flavins or benzyl viologen as electron donors for reduction of nitrate to nitrite. NADH-dependent nitrate reductase is most prevalent in plants. Hence, NR activity in plants can be measured by following the oxidation of NAD (P) H at 340nm. However, NR activity is commonly measured by colorimetric determination of nitrite produced.

Materials

- **Potassium Phosphate Buffer** 0.1M (pH 7.5)
- **Potassium Nitrate 0.1M**

  Dissolve 1.01g potassium nitrate in 100ml water.
- **NADH 2mM.**

  Dissolve 14mg NADH disodium salt in 10ml water.
- Sulphanilamide 1% (w/v):

  Dissolve 1g sulphanilamide in 100ml 2.4N hydrochloric acid.
- N-(1-naphthyl) Ethylenediamine Dihydrochloride:

  0.02%. Dissolve 20mg in 100Ml water.
- Potassium Nitrite Standard Solution (0.01M):-

  Dissolve 851mg pure potassium nitrite in 100mL water in a standard flask.

  Dilute 10mL of this solution to 100mL and use as working standard solution.

Enzyme Extract

Homogenize a weighed quantity of the plant material in a known volume of medium (6mL for 1g fresh tissue) containing 1mm EDTA, 1-25mm cysteine and

25mm potassium phosphate adjusted to a final pH 8.8 with KOH. Filter through four layers of cheese cloth and centrifuge for 15 min at 30,000g. Decant the supernatant through glass wool and use for assays. Extract under ice-cold conditions.

Procedure

1. Pipette out 0.5ml of phosphate buffer (pH 7.5) in a test tube.
2. Add 0.2 ml of potassium nitrate solution
3. Initiate the reaction by addition of 0.2ml of enzyme extract. Set up a control in the same way but with water instead of enzyme extract.
4. Incubate at 30 °C for 15 minutes.
5. Terminate the reaction by the rapid addition of 1ml sulphanilamide followed by 1ml naphthyl ethylenediamine reagent.
6. Wait for 30 minutes.
7. Measure the absorbance at 540nm.
8. Prepare a standard graph with sodium nitrite. Pipette out different known aliquots of potassium nitrite standard solution into a series of test tubes and make up the volume in each tube to 2ml by adding water.
9. Repeat the step 5 to 7.

Calculation

Activity is expressed as micromole nitrite produced per min per mg protein (or per g fresh tissue).

Notes

1. The enzyme reaction rate is linear over a 30 min period.
2. The pink color produced by nitrite is stable for 2-3h.

5.4 Catalase

Catalase has a double function as it catalases the following reaction:

Decomposition of hydrogen peroxide to give water and oxygen

$$2H_2O_2 \xrightarrow{Catalase} 2H_2O + O_2$$

Oxidation of H donors for example methanol, formic acid, phenol with the consumption of one mole of peroxide.

$$ROOH + AH_2 \xrightarrow{Catalase} H_2O + ROH + A$$

Principle

TheUV light absorption of hydrogen peroxide solution can be easily measured between 230 and 250nm. On decomposition of hydrogen peroxide by catalse, the absorption decreases with time. The enzyme activity could be arrived at from this decrease. But this method is applicable only with enzyme solutions which do not absorb strongly at 230 to 250nm.

Materials

- **Phosphate buffer,** (0.067M (pH 7.0)

  Dissolve 3.522g KH_2PO_4 and 7.268g Na_2HPO_4. $2H_2O$ in distilled water and make up the volume to one liter

- **Hydrogen peroxide-phosphate buffer:**

  Dilute 0.16ml of H_2O_2 (10% w/v) to 100ml with phosphate buffer. Prepare fresh. The absorbance of the solution should be about 0.5 at 240nm with a 1 cm light path.

- **Enzyme extract:**

  Homogenize plant tissue in a blender with M/150 phosphate buffer (assay buffer diluted 10 times) at 1-4° C and centrifuge. Stir the sediment with cold phosphate buffer; allow standing in the cold with occasional shaking and then repeating the extract once or twice. The extract should not be takes longer than 24 hours. Use the combined supernatant for the assay. The catalase activity can change considerably on storage of tissue. In comparative studies, therefore, always use the same conditions of the tissue, storage and temperature.

Procedure

1. Set the wavelength at 240nm
2. Final volume of the reaction mixture should be 3ml approximately
3. Read against a control cuvette containing enzyme solution as in the experimental cuvette, but containing H_2O_2-free PO_4 buffer (M/15).
4. Pipette into the experimental cuvette

5 Add 3ml of H_2O_2-PO_2 buffer. Mix in 0.01 to 0.04 ml sample with the glass or plastic rod flattened at one end. Note the time Δt require for a decrease in absorbance from 0.45 to 0.4. This value is used for calculations. It t is greater than 60 sec repeat the measurements with a more concentrated solution of the sample.

Calculation

One g tissue is homogenized in a total volume of 20ml, dilute 1 to 10 volumes with water and take 0.01ml for assay. The absorbance at 240nm decreased from 0.45 to 0.4 in 13.9 sec.

$$= \frac{17}{13.9} = 1.22 \text{ units in the assay mixture (or)}$$

$$= \frac{1.22 \times 10}{0.01} = 1220 \text{ units/ml extract i.e., } 2.44 \times 10^4 \text{units/g tissue}$$

Calculate the concentration of H_2O_2 using the extinction coefficient 0.036 per µmole per ml.

5.5 Peroxidase

Peroxidase (POD) includes in its widest sense a group of specific enzymes such as

NAD- Peroxidase, NADP- Peroxidase, fatty acid peroxidase etc. as well as group of very non specific enzymes from different sources which are simply known as POD. POD catalyses the dehydrogenation of a large number of organic compounds such as phenols, aromatic amines, hydroquinones etc. POD occurs in animals, higher plants and other plants and other organisms.

Principle

Gluaiacol is used as substrate for the assay of peroxidase.

$$\text{Guaiacol} + H_2O_2 \rightarrow \text{Oxidized guiacol} + 2H_2O_2$$

The resulting oxidized (dehydrogenated) guaicol is probably more than one compound and depends on the reaction conditions. The rate of formation of guaiacol dehydrogenation product is measure of the POD activity and can be assayed spectrophotometrically at 436nm.

Materials

- **Phosphate buffer**: 0.1 M (pH 7.0)
- **Guaiacol Solution**: (20mM)

  Dissolve 240mg of guaiacol in water and make up the volume to 100ml. it can be stored in many months.
- Hydrogen peroxide solution (0.042% = 12.3mM)

Dilute 0.14ml of 30% H_2O_2 to 100 ml of water. The extinction of this solution should be 0.485 at 240nm. Prepare freshly.

- Enzyme extract:

Extract1g of fresh plant tissue in 3ml of 0.1M phosphate buffer pH 7 by grinding with a pre-cooled mortar and pestle. Centrifuge the homogenate at 18,000g at 5°C for 15 min. use the supernatant as enzyme source within 2- 4 hours. Store the on ice till the assay is carried out.

Procedure

1. Pipette out 3ml buffer solution, 0.05ml guaiacol solution, 0.1 ml enzyme extract and 0.03 ml hydrogen peroxide solution in a cuvette.

   (Bring the buffer solution to 25°C before assay)
2. Mix well and place the cuvatte in the spectrophotometer.
3. Wait until the absorbance has increased by 0.05. Start a stop-watch and note time required in minutes (Δt) to increase the absorbance by 0.1.

Calculation

Since the extinction coefficient of Guaiacol dehydrogenation product at 436nm under the conditions specified is 6.39 per micromole, the enzyme activity per liter of extract is calculated as below.

$$\text{Enzyme activity units/liter} = \frac{3.18 \times 0.1 \times 1000}{6.39 \times 1 \times \Delta t \times 0.1} = \frac{500}{\Delta t}$$

Notes

Most accurate values are obtained when Δt is between 1 to3 min. the enzyme extract has therefore to be diluted appropriately.

Or

If the substrate is O- Dianisidine

Procedure

1. The oxidized o-dianisidine (yellow orange colored compound) is measured at 430nm.
2. Take 3.5 ml phosphate buffer (pH 6.5) in a clean dry cuvette.
3. Add 0.2 ml enzyme extract and then place the cuvette in the spectrophotometer set at 430nm.
4. Add 0.2ml 0.2M H_2O_2 and mix. Immediately start the stop watch.
5. Read the initial absorbance and them at every 30 sec interval up to 3 min.
6. (If the rate of increase is very high, repeat the assay with dilute extracts. plot increase in absorbance against time.)
7. From the linear phase, read the change in absorbance per min.
8. Express enzyme activity in terms of rate of increased absorbance per unit per mg protein or tissue weight.
9. Water blank is used in the assay.

5.6 Polyphenol Oxidase

Phenol oxidases are copper proteins of wide occurrence in nature which catalyse the aerobic oxidation of certain Phenolic substances to quinines which are autoxidized to dark brown pigment generally known as melanin's. Theses enzymes are assumed to be a single enzyme with broad specificity although there is some evidence for the presences of more than one phenol oxidase in the certain tissue. Each individual enzyme tends to catalyze the oxidation of one particular phenol or Phenolic compound more readily than the other. The phenol oxidase (POD) comprises of catechol oxidase and laccase. The activities of these enzymes are more important with regards to first plant defiance mechanism against pest and disease and second appearance palatability and use of plant products. Fresh fruits, vegetables, mushroom etc. contains these enzymes considerably.

Principle

The intensely 2- nitro-5 – thiobenzoic acid (TNB) with an absorption maximum at 412nm react with the quinines generated through the enzymatic oxidation of 4- methylcatechol (catechol oxidase) and 1,4- dihydroxybenzen (laccase) to yield colorless adducts. The decrease in the absorbance of the yellow- color due to enzyme activity is measured.

Materials

- **Citrate phosphate buffer** (0.2M pH 6.0)
- **Preparation of 2 nitro- 5- thiobenzoic anion** (TNB):

Add30mg of sodium borohydride to a suspension of Ellman's reagent, i. e. 5, 5 disulphide is quantitative reduction to the intensely yellow, water-soluble thiol. This solution is stable for at least one week when stored at 4°C.

- **Preparation of the quinine solution:**

  Dissolve4-methyl-1, 2- benzoquinone in a double distilled water in a 50 ml volumetric flask by bubbling nitrogen gas until the quninone is completely dissolved.

  Prepare *p*-benzoquinonc solution also in a similar manner. Both solutions are stable for 30 min, a time sufficient to carry out the spectrophotometric assay.

- **Substrate solution:**

  4-methylcatechol (2mm) for the catechol oxidase assay quininol (1,4-dihydroxybenzene, 2mm) for laccase assay.

- **Enzyme extract:**

  Prepare first acetone powder of fresh plant tissue by blender homogenizing 25 g tissue in two successive 100ml aliquots of cold acetone. The homogenate was collected by Buchner filtration through whatman No.1 filetr paper following both grindings. The homogenate was air dried until free of acetone odour, the resulting dry powder was weighed and freezer-stored in cold containers.

  To get the crude enzyme preparation, mix 100mg acetone powder with 2.5ml of 0.2M citrate phosphate buffer (pH 6.0). 1ml of 1% triton-x -100, 6.5 of water and 500mg polyamide. Shake for 1 hours and filter. Use filtrate as enzyme source.

Procedure

1. Pipette out into clean 1cm cuvette of 1.4 ml citrate 0.1 M phosphate buffer (pH 6.0), 0.5ml of TNB and 1 ml substrate solution.
2. Start the reaction by adding 0.1 ml of enzyme preparation and immediately note down the absorbance at 412nm in a spectrophotometer already set.
3. Follow the decrease in absorbance at 30sec intervals and record.

Or

Alternative Methods for Polyphenl Oxidase

1. Add 2.5ml of 0.1 M phosphate buffer pH 6.5, 0.3ml of acetone solution (0.01M) into cuvette and set the spectrophotometer at 495nm.
2. Now add 0.2ml of enzyme extract and start recording the change in absorbance for every 30sec upto 5min.
3. For the above procedure, the enzyme extract may be prepared by grinding 5g leaves with a mortal and pistal in about 20 ml medium containing 50Mm Tris- HCl, pH 7.2, 0.4 M sorbitol and 10 Mm NaCl.
4. Centrifuge the homogenate at 2000 rpm for 10 minutes.

5. Collect the supernatant and use it for the assay.

Calculation

Read the change in absorbance per minute from the linear part of the curve. Calculate the enzyme units according to the following problem.

$$\text{Units of the test enzyme} = K \times (\Delta A/\text{min})$$

Whereas,

K is 0.272 for Catechol Oxidaase and 0.242 for Laccase.

One unit of either Catechol Oxidase or Laccase is defined as the enzyme which transforms 1 μmole of dehydration to 1μmol of quinine per min under the assay condition. One unit is equivalent to the consumption of 1μmole of TNB.

5.7 Amino Transferase

Amino transferases are enzymes catalyzing the transfer of an amino acid group plus a proton and an electron pair, from an amino donor compound to the carboxyl position of an amino accepter compound. Usually an amino acid acts as the amino donor and a 2-oxo-acid as the amino accepter. In same case aldehyde may serves as amino accepters, and amines act as donors. The enzymatic transfer of amino groups play an important role in many metabolic processes where the interconversion of nitrogen containing molecules is involved. Nitrogen, following its initial assimilation in to glutamine and glutamate can be distributed to many other compounds by the action of aminotrasferases. Glutamate is often the amino-donor substrate in biosynthetic transamination reaction. This reaction regenerates 2-oxyglutarate for necessary ammonia assimilation through GDH and the glutamine synthetase/glutamate synthase routes.

Principle

Glutamateoxaloacetate aminotransferase (GOT) catalyses the reversible interconversions between glutamate and aspartate and their 2-oxo analogues.

$$\text{Glutamate} + \text{Oxaloacetate} \xrightleftharpoons{\text{GOT}} \text{2–Oxoglutarate} + \text{Aspartate}$$

The oxaloacetic acid is measured colorimatrically by a reaction with 2,4-dintrophythydrazine giving a brown – colored hydrazone after the addition of 0.4N sodium hydroxide.

Materials

- **Phosphate Buffer (*Ph7.4*)**

  Add a11.3g of dry anhydrous disodium hydrogen phosphate and 2.7 g dry anhydrous potassium dihydrogen phosphate in 1liter and make up to the mark with water. Check the pH and store at 4°C.

- **Substrate Solution:**

  Dissolve 13.3 g DL- aspartic acid in minimum amount of 1N sodium hydroxide and prepare a solution with pH 7.4 (about 90ml is required). Add 0.146g of 2-Oxoglutarate and dissolve it in little more sodium hydroxide solution, adjust the pH 7.4 and then make to 500ml with phosphate buffer.

Divide into 10ml portions and store frozen at -15°C.

- **Pyurate Standard:**

  Dissolve 22mg sodium pyruvate in 100ml od distilled water in a standard flask.

- **2,4-dinitrophynylhydrazine (DNPH):**

  Dissolve 19.8mg of dinitrophylhydrazune in 100ml conc. HCl and make up to 100ml with distilled water. Store it amber colored bottle at room temperature.

- **Sodium hydroxide (0.4N):**

  Dissolve 16g of sodium hydroxide in 1 liter of water.

- **Enzyme extract:**

  Prepare the crude extract by grinding the plant tissue in 0.2M Potassium phosphate Ph7.5 in a homogenizer for 2 minutes. Pass the slurry through eight layers of cheesecloth and then centrifuge at 25000 rpm for 15 minute to get the enzyme fraction.

Procedure

1. Warm 0.5ml of substrate solution a in water bath at 37°C for 3 minutes
2. Add a 0.2 ml of enzyme extract and mix gently.
3. Incubate for a 60 minutea at 37°C.
4. Remove the tubes form the bath and immediately add 0.5ml dinitrophynylhydrazine solution and mix well.
5. Mix 0.5 ml of substrate with 0.5 of DNPH solution and then add 0.1ml enzyme extract for control.
6. Allow the DNPH to react for 20 minutes at room temperature.
7. Add 5 ml of 0.4N Sodium hydroxide, mix well and leave for 10 minutes.
8. Record the absorbance at 510nm.
9. Pipette out pyruvate standard 0.05 to 2.00ml and make up to 0.2 ml. Add 0.5ml substrate and 0.5ml DNPH solution. For blank mix 0.5 ml substrate, 0.2ml water and 0.5ml DNPH solution.
10. Then repeat step 6 to 8.

Calculations

Thepyruvate formed by enzyme is responsible for the absorbance difference between test and control. The pyruvate in standard produce the difference between standard and blank. Express the enzyme activity as μmole of pyruvate formed per minutes per mg of protein.

5.8 Lipase

Lipase hydrolyses triglycerides to release free fatty acids and glycerol.

$$\text{Triglycerides} + H_2O \rightarrow \text{Glycerol} + \text{fatty acids}$$

During germination of oilseeds, lipases play an important role in hydrolyzing the stored oils so that the required energy for growth and carbon skeleton for synthesis of new compounds are produced are good sources of lipases.

Principle

Thequantity of fatty acid released in unit time is measured by the quantity of NaOH required to maintain pH constant. The Millie equivalent of alkaline consumed is taken as a measure of the activity of the enzyme.

Materials

- **Substrate:**

  Take 2ml of any clear vegetable oil, neutralize to pH 7.0, if necessary, and stir well with 25ml of water in the presence of 100mg bile salts (sodium taurocholate) till an emulsion is formed. Addition of 2g gum Arabic hastens emulsification.

- **0.1 N NaOH :**

  0.4 g of sodium hydroxide in 100 ml of water

- **Enzyme extraction:**

  Grind of known quantities of samples with the mortal and pestle. Homogenize the tissues with twice the volume of ice-cold acetone. Filter and wash the powder successively with acetone, acetone: ether (1:1) and ether. Air dries the powder. This acetone powder can be stored in a refrigerator. Extract 1g of the powder in 20ml ice-cold water or a suitable buffer. Centrifuge at 15000 rpm for 10 minutes and use the supernatant as enzyme source.

- **50Mm phosphate buffer (pH 7.0)**

Procedure

1. Take 20ml of substrate in 500ml beaker. Add 5ml of phosphate buffer (pH 7.0)
2. Set the beaker on top of a magnetic stirrer cum hot plate and stir the contents slowly. Maintain the temperature at 35°C. Dip the electrodes of a Ph meter in the reaction mixture. Note the Ph and adjust it to 7.0.
3. Add enzyme extract (0.5ml), immediately record the Ph and set the timer on. Let it be pH at zero time.
4. At frequent intervals (10 min) or as the pH drops by about 0.2 unit add 0.1N NaOH to bring pH to the initial value. Continue the titration for 30 – 60 min period.
5. Note the volume of alkali consumed.

Calculation

The enzyme activity is defined as the amount of enzyme which releases one milliequivalent of free fatty acid per minute per g of sample. Specific activity is expressed as milliequivalents/min/mg of protein.

Activity meq/min/g of sample

$$= \frac{\text{Volume of alkali consumed} \times \text{Strength of alkali}}{\text{Wt. of sample in g} \times \text{Time in min}}$$

5.9 Acid Phosphatase

Phosphatase are the enzymes which librate inorganic phosphate from organic phosphate esters of primary and secondary alcohols, sugar alcohols, phenols and amines. There are two major classes of phosphatases, alkaline phosphatase and acid phosphatase, which refer to pH optima of these enzymes.

Principle

Theenzyme phosphates hydrolyzes *p*-nitrophynol phosphate, the released *p*-nitrophenol is yellow in color in alkaline medium and is measured at 405nm. The optimum pH for acid and alkaline phosphatases are 5.3 and 10.5respectively.

Methods and Materials

1. Acid phosphatase:-

Materials

- **Sodium hydroxide (0.085N):**

  Dissolve 0.85g of sodium hydroxide in 250 ml of distilled water

- **Substrate solution:**

  Dissolve 1.49g of EDTA, 0.84g of citric acid and 0.03g of *p*-nitrophenyl phosphate in 100ml of water and adjust to Ph 5.3.

- **Standard preparation:**

  Weigh the 69.75mg of *p*-nitrophenyl and dissolve in 5.0ml of distilled water (100mM)

- **Enzyme extract:**

  Homogenize 1g of fresh tissue in 10ml of ice-cold 50Mm Citrate buffer (pH5.3) in a pre-chilled pestle and mortar. Filter through four layer of cheese cloth. Centrifuge the filtrate at 10,000 rpm for 10 min. use the supernatant as a enzyme source.

Procedure

1. Incubate 3ml of substrate solution at 37°C for 5 min.
2. Add 0.5 ml of enzyme extract and mix it well.
3. From the above mixture remove immediately 0.05ml and mix it with 9.5ml of sodium hydroxide solution 0.085N. This corresponds to zero time assay.
4. Incubate the remaining solution (substrate + enzyme) for 15 min at 37°C
5. Pipette out 0.5 ml of solution and mix it with 9.5ml of of sodium hydroxide solution 0.085N.
6. Measure the absorbance of blank at 405nm.

7. Take 0.2 ml to 1.0 ml of (4 to 20 mm) of the standard, dilute to 10.0 ml with sodium hydroxide solution. Read the color and standard curve.

Calculation

Specific activity is expressed in *m* moles of *p*-nitrophenol release per minper mg of protein.

2 Alkaline phosphatse

For the alkaline phosphates extract the enzyme in 50Mm glycine NaOH buffer pH 10.4. its optimum function is at pH 10.5. the assay procedure is similar to that for acid phosphatase, except for the substrate solution. Prepare the substrate solution as follows.

Dissolve 375mg of glycine, 10mg of magnesium chloride, 165mg *p*-nitrophenyl phosphate in 42ml of 0.1N Sodium hydroxide and dilute to 100ml. adjust the pH10.5.

5.10 Amylases

Amylase is a hydrolytic enzyme which breaks down many polysaccharides e.g., starch, which is a polymer of glucose units linked by alpha 1-4 bond, to yield the disaccharide (maltose) as the end product.

$$\text{Starch} + nH_2O \xrightarrow{\text{Amylase}} \text{Maltose}$$

Principle

The reducing sugar produced by the action of alph and beta amylase reacts with dinitrosalicylic acid and reduce it to a brown colored product.

Materials

- **Sodium citrate buffer (0.1M pH 4.7)**
- **Starch 1% solution**

  Prepare a fresh solution of dissolving 1g starch in 100ml of acetate buffer. Slightly warm the starch solution, if necessary.
- **Dinitrosalicylic Acid Reagent**
- **40% Rochelle Salt solution** (Potassium Sodium Tartarate)
- **Maltose solution:**

  Dissolve 50mg of maltose in 50ml of distilled water in a standard flask and store it in a refrigerator.
- **Extraction of Amylase**

  Extract 1 g of sample materials with 5 to 10 volume of ice cold 10 Mm calcium chloride solutions overnight at 4°C or for 3h at room temperature. Centrifuge the extract at 50,000 rpm for 20 minutes. The supernatant is used as an enzyme source.
- **Extraction of beta amylase (free and bound)**

  The free beta amylase is extracted from acetone defatted sample material in 66mm phosphate buffer (Ph7.0) containing 0.5M NaCl. The extract is centrifuged at 20,000rpm for 15 min. the supernatant is used as an enzyme source.

Procedure

1. Pipette out 1ml of starch solution and1ml of properly diluted enzyme in a test tube.
2. Incubate it at 27°C for 15 min.
3. Stop the reaction by adding 2 ml of dinitrosalicylic acid reagent.
4. Heat the solution in a boiling water bath for a 15 min.
5. While the tubes are warm, add 1ml potassium sodium tartrate solution.
6. Then cool it in running tap water.
7. Make up the volume to 10ml by addition of 6ml of distilled water.
8. Read the absorbance at 560nm
9. Terminate the reaction at zero time in control tubes.
10. Prepare a standard graph with 0- 100 μg of maltose.

Calculations

The unit of alpha and beta amylase is expressed as mg of maltose produced during 5 min incubation with 1% starch.

Notes

The extraction procedure given is suitable for the cereal grains. There are varieties of extraction procedures used for this purpose depending on the source materials.

5.11 Cellulose

Cellulose, a major structural polysaccharide in plants, is the most abundant organic compound in nature, and is composed of glucose units joined together in the form of the repeating unit's of the disaccharide cellobiose with numerous cross linkages. It is also a major component in many of the farm wastes.

Principle

Cellulose undergoes acetolysis with acetic/nitric reagent forming acetylated cellodextrins which get dissolved and hydrolyzed to form glucose molecules on treatment with 67% H_2SO_4. This glucose molecule is dehydrated to form hydroxymethyl furfural which forms green colored product with anthrone and the color intensity is measured at 630 nm.

Materials

- Acetic/Nitric reagent:

  Mix 150 ml of 80% acetic acid and 15 ml of concentrated nitric acid.

- Anthrone reagent:

  Dissolve 200 mg anthrone in 100 ml concentrated sulphuric acid. Prepare fresh and chill for 2 h before use.

- 67% Sulphuric acid.

Procedure

1. Add 3 ml acetic/nitric reagent to a known amount (0.5 g or 1 g) of the sample in a test tube and mix in a vortex mixer.
2. Place the tube in a water-bath at 100°C for 30 min.
3. Cool and then centrifuge the contents for 15–20 min.
4. Discard the supernatant.
5. Wash the residue with distilled water.
6. Add 10 ml of 67% sulphuric acid and allow it to stand for 1 h.
7. Dilute 1 ml of the above solution to 100 ml.
8. To 1 ml of this diluted solution, add 10 ml of anthrone reagent and mix well.
9. Heat the tubes in a boiling water-bath for 10 min.
10. Cool and measure the color at 630 nm.
11. Set a blank with anthrone reagent and distilled water.
12. Take 100 mg cellulose in a test tube and proceed from Step No. 6 for standard.
13. (Instead of just taking 1 ml of the diluted solution (Step 7) take a series of volumes (say 0.4–2 ml corresponding to 40–200 μg of cellulose) and develop the color).

Calculation

Draw the standard graph and calculate the amount of cellulose in the sample.

5.11 Sucrose Synthase

Sucrose synthase(EC2.4.1.13) is anenzymethatcatalyzesthechemical reactionNDP-glucose + D-fructose ⇌ NDP + sucrose. Thus, the twosubstratesof this enzyme areNDP-glucoseandD-fructose, whereas its twoproductsareNDPandsucrose. This enzyme belongs to the family ofglycosyltransferases, specifically the hexosyltransferases. The systematic name of this enzyme class isNDP-glucose:D-fructose 2-alpha-D-glucosyltransferase. Other names in common use includeUDPglucose-fructose glucosyltransferase,sucrose synthetase,sucrose-UDP glucosyltransferase,sucrose-uridine diphosphate glucosyltransferase, anduridine diphosphoglucose-fructose glucosyltransferase. This enzyme participates instarch and sucrose metabolism.

Materials

Preparation of Sucrose Synthase

All steps were carried out at 4°C. Nodules (10 g) were harvested 40 d after planting and homogenized with a mortar and pestle in 20 ml of 10 mm Kphosphate(pH 7.2) containing 1 mm EDTA and 5 mm 2-mercaptoethanol (buffer A). A suspension of 10 g of insoluble PVP in 20 ml of buffer A was added to the homogenate, the mixture filtered, and the filtrate centrifuged at 30,000g for 15 min. The supernatant was fractionated by the addition of saturated(NH.4)SO4 solution, and the fraction which precipitated between 30 and 50% saturation collected by centrifugation at 30,000g for 10 min. The precipitate was dissolved in 2.5 ml of buffer A and applied to a Sephadex G-200 column (2.5 x 85 cm) which had been previously equilibrated with 25 mm Kphosphate (pH 7.2) containing 50 mm KCI, 1 mm EDTA, and 5 mm 2-mercaptoethanol. Fractions of 5 ml were collected and those which contained sucrose synthase activity were pooled, dialyzed against buffer A and applied to a DEAE-cellulose column 1.5 x 10 cm) which had been previously equilibrated with buffer A. The column was washed with buffer A until the A at 280 nm was less than 0.01 and eluted with a gradient produced by introducing 100 ml of 0.5 M KCI in buffer A into 100 ml of buffer A. Fractions of 4 ml were collected. Active fractions were pooled and dialyzed against 50 mm Hepes-KOH buffer (pH 8.5) containing 5 mm sucrose, 10 mm MgC92, and 5 mm 2-mercaptoethanol (buffer B). The preparation was applied to a PBA-60 affinity column (1.5 x 5 cm) which had been washed with 250ml of buffer B containing 200 mm sucrose, followed by 25 ml of buffer B. After unbound protein had been removed, the enzyme was eluted with 0.1 M Tris-HCl buffer (pH 8.5) containing 5 mM2-mercaptoethanol. Fractions of 3 ml were collected and those which contained activity were pooled and dialyzed against 20mM K-phosphate (pH 7.0) containing 5 mM 2-mercaptoethanol.

Assay of Sucrose Synthase Activity

All assays were carried out at 300C. Activity in the sucrose cleavage direction was assayed by three methods (assays A-C). Three other methods were used to assay synthesis activity (assays D-F).Assay A. The production of UDPglucose was coupled to the reduction of NAD in the presence of excess UDP glucose dehydrogenase and the change in A at 340 nm followed. Reaction mixtures contained in a volume of I ml, 20 umol Hepes-KOH buffer (pH 7.5), 100, umol sucrose, 2, umol UDP, 1.5 Mmol NAD,25 ,g UDPglucose dehydrogenase, and an appropriate volume of enzyme.

Assay A was used to study the kinetic parameters and the effects of pH, fructose, metabolites, and salts on the cleavage reaction.

Assay B.: The effect of inhibitors on sucrose cleavage was studied in reaction mixtures which contained, in a volume of 1ml, 20, mol Hepes-KOH buffer (pH 7.5), 100, mol sucrose, 2 Amol UDP, and an appropriate volume of enzyme. The reaction was stopped after 30 min by heating in a boiling water bath for 2 min, and fructose was determined by the reducing sugar method of Nelson (16).

Assay C Sucrose cleavage activity with glucosyl acceptors other than UDP, and the inhibition of sucrose cleavage byUDPglucose, were studied in reaction mixtures which contained,in a volume of1 ml, 20 ,mol Hepes-KOH buffer (pH 7.5), 100Mmol sucrose, 2 Mmol nucleotidc diphosphate, and an appropriate volume of enzyme. The reaction was stopped after 30 min by heating in boiling water bath for 2 min. Fructose was determined from the change in A at 340 nm following the addition of a solution which contained, in a volume of 0.2 ml, 70,umolTris-HCI buffer (pH 7.5), 60,umol KCI, 2 umol $MgCl_2$, 0.36umol NADP, 1.2,umol ATP, 20 ,ug hexokinase, 20,g P-glucoseisomerase, and10,ug glucose-6-P dehydrogenase.

Assay D Reaction mixtures contained, in a volume of1 ml,20,mol Hepes-KOH buffer (pH 7.5), 15,mol fructose, 2 Mmol UDPglucose, and an appropriate volume of enzyme. The reaction was stopped after 30 min by heating for 2 min in a boiling water bath. UDP was determined from the change in A at 340nm following the addition of a solution which contained, in a volume of 0.2 ml, 5 Mmol MgC12, 0.4 gmol P-enolpyruvate, 0.15,umol NADH and 20,umol KCI, 25 Mg pyruvate kinase, and 25Mg lactate dehydrogenase. This assay was used to study the effects of pH, sucrose, metabolites, salts, and inhibitors on the synthesis reaction. When the effect of pH was studied, 0.12 mmol Hepes-KOH buffer (pH 7.5) was included in the solution added after the reaction was stopped.

Assay E Activity was assayed by coupling the production of nucleotide diphosphate to the oxidation of NADH in the presence of excess pyruvate kinase and lactate dehydrogenase. Reaction mixtures contained, in a final volume of1 ml, 20umolHepes-KOH buffer (pH 7.5), 1u5 mol fructose,,u2 mol UDPglucose,5 Umol MgCI2, 0.4 mol P-enolpyruvate, 0.15 Mmol

NADH, ,2u0 mol KCI, 25 MAg pyruvate kinase, 25 MAg lactate dehydrogenase, and an appropriate volume of enzyme. The decrease in A at 340 nm was followed. The effect of ADPglucosewas studied in reaction mixtures of the same composition exceptthat UDPglucose was omitted and the concentration of ADPglucose varied.

Assay F Reaction mixtures were of the composition describedfor assay D. The reaction was stopped after 30 min by the addition of0.1 ml of0.1M citrate buffer (pH 5.0) and heating in a boiling water bath for 2min. Glucose was determined according to the method of Blakeney and Matheson (5) following the addition of0.1 mg of in vertase. Assay F was used to study the inhibition of sucrose synthesis by UDP and ADP. One U of activity is defined as the amount of enzyme which catalyzed the formation I of umol of product/mmn. Protein was determined by the Folin-Lowry method.

Analysis of Data

The kinetic constants for UDP, UDPglucose,sucrose, and fructose were determined by fitting the data. The equations used were:

$$V = VA/\ K_{ia}+K_{h}A+\ K_{h}B+AB \qquad \text{.......(1)}$$

$$V = VA/\ K_{a}+A+\ A/K_{i} \qquad \text{.......(2)}$$

Kinetic constants for ADP, CDP, and ADPglucose were calculated by nonlinear regression analysis of initial velocity data.

Gel Electrophoresis. Sucrose synthase activity was located by incubating gels at 37°C for 30 min in a reaction mixture which contained, in a volume of 10 ml, 0.2 mmol Hepes-KOH buffer (pH 7.5), 1 mmol sucrose, and 20 umolUDP. The gels were rinsed in distilled II_20 and incubated in IN NaOH containing 1% (w/v) triphenyl tetrazolium chloride for10 min at 37°C. The mol wt of sucrose synthase was estimated by PAGE using thyroglobulin, ferritin, catalase, lactate dehydrogenase, and BSA as standards. SDS-PAGE gel electrophoresis was carried out as described by Weber and Osbom (30). Phosphorylase b, BSA,ovalbumin, carbonic anhydrase, trypsin inhibitor, and a-lactalbumin were used as standards.

References

Bergmeyer H. U and Bernt E. (1974) In: Methods of Enzymatic Analysis (Ed Bergmeyer) Vol 2 p. 735.

Chelikani P, Fita I, Loewen PC (January 2004). "Diversity of structures and properties among catalases". Cell. Mol. Life Sci. 61 (2):

Delma Dohertry (1970) In: Methods in enzymol vol.17 Part A (Eds Tabor, H., and Tabor, C.W.,) p. 850.

DOI: https://doi.org/10.1104/pp.110.154203

Fiske C H. and Subba Row Y (1925) J BiolChem66 575.

Gnanam, A. and Francis, K. (1976) Plant and animal Biochem. J. 3, p.11.

Hageman, R H and Reed, A J (1980) In:Methods in Enzymology.Vol. 69 Part C(EdAnthony San Pietro) Academic Press New York p 270.

http://biocyclopedia.com/index/plant_protocols/enzymes/sucrose_synthase.php

http://docsdrive.com/pdfs/ansinet/jas/2002/474-484.pdf

http://nfscfaculty.tamu.edu/talcott/courses/FSTC605/Papers%20Reviewed/Review-PPO.pdf

http://shodhganga.inflibnet.ac.in/bitstream/10603/74321/12/12_chapter%202.pdf

http://thescipub.com/PDF/ajbbsp.2011.141.145.pdf

http://users.auth.gr/palexios/Publications/Pdf-Chapters/Catalases%20in%20Plants.pdf

http://www.pomics.com/agoreyo_3_3_2010_66_69.pdf

http://www.sfu.ca/bisc/bisc367/handouts/lab05.pdf

http://www.uky.edu/~jrawls/bio410/LAB%20MANUALnri.pdf

http://www.uvm.edu/~bio1and2/lab/Lab%20manuals%20Fall%202011/Enzyme%20Activity%20Phosphatase.pdf

https://doi.org/10.1016/0003-9861(56)90100-X

https://doi.org/10.1016/0005-2744(69)90127-2

https://doi.org/10.1016/S0168-9452(00)00224-7

https://doi.org/10.1104/pp.60.3.379

https://doi.org/10.1186/s13007-015-0090-6

https://doi.org/10.21769/BioProtoc.1213

https://doi.org/10.21769/BioProtoc.1444

https://doi.org/10.21769/BioProtoc.1978

https://www.ncbi.nlm.nih.gov/pmc/articles/PMC1066016/pdf/plntphys00558-0023.pdf

https://www.ncbi.nlm.nih.gov/pmc/articles/PMC1077577/pdf/plntphys00815-0291.pdf

Jayaraman J (1981) In: Laboratory Manual in Biochemistry Wiley Eastern Limited New Delhi p133.

Kalaivani M., Jebanesan A., Maragathavalli S., Annadurai B. & Gangwar S. K. (2013). G.J.B.B., VOL.2 (2): 267-277.

King E J (1932) BiochemJ26 292.

Kruger J E (1972) Cereal Chem 49 379.

Luck H (1974) In: Method in EnzymaticAnalysis2 (Ed Bergmeyer) Academic Press New York p 885.

Malik C P and Singh M B (1980) In: Plant Enzymology and Histoenzymology. Kalyani Publishers New Delhi p 53.

Matrhew Morell and Les Copeland (1984).Sucrose Synthase of Soybean Nodules. Plant Physiol. (1985) 78, 149-154.

Niku-Paavol, M L., Nummi M., Kanchkun A., Daussant J and Enari T M. (1972) Cereal Chem49 580.

Putter J (1974) In: Methods of enzymatic analysis2 (Ed Bergemyer) Acadenic Press New York p 685.

Sadasivam, S and Gowri, G (1981)Experientia37552.

Snell, F D and Snell, C T (1949) Colorimetric Methods of Analysis2 Van Nostrand Co Inc New Jersey p 805.

Splittstoessetr W. E., Chu M C, Steward S. A and Splittstoesser S. A. (1976) Plant Cell Physiol17 83.

Updegroff D M (1969) Anal Biochem 32 420.

6

Chapter

Assay for Antioxidants

6.1 Assays for Enzymatic Antioxidants

Enzyme Antioxidants

Despite aerobic organisms generating Reactive oxygen species ROS as a natural byproduct of O_2 metabolism, metazoan have evolved mechanisms of defense that cope with this reactive chemical species. However, oxidative stress occurs in situations of imbalance. This is when ROS levels increase and protective compounds, namely antioxidants, are overwhelmed. As a result, significant damage to cell structures like lipids, proteins and DNA is exerted. Thus, O_2 is considered a potential mutagen, clastogen and teratogen that may be responsible for, or at least part of, the background genetic instability. The protective effect of antioxidants is universally accepted, either inherent in the body or ingested. However, certain aspects such as the mechanisms of action, undefined properties of compounds or compounds with dual behavior (pro oxidant and antioxidant) remain unclear and are subjected to investigation. For instance, the elucidation of how antioxidant properties operate in vitro can provide a better understanding of the, sometimes confounding, in vivo situation.

In the present manual, several of the most well-known antioxidants are considered. These include: superoxide dismutase, catalase, peroxidase, glutathione S-transferase and polyphenol oxidase were analyzed.

6.1.1 Assay of Superoxide Dismutase (SOD)

SOD was assayed according to the method of Kakkar *et al.* (1984).

Principle

The assay of SOD is based on the inhibition of the formation of NADH-phenazine methosulphate-nitroblue tetrazolium formazon. The color formed at the end of the reaction can be extracted into butanol and measured at 560nm.

Materials

- Sodium pyrophosphate buffer (0.025M, pH 8.3)
- Phenazine methosulphate (PMS) (186μM)
- Nitroblue tetrazolium (NBT) (300μM)
- NADH (780μM)
- Glacial acetic acid

- n-butanol
- Potassium phosphate buffer (50mM, pH 6.4)

Procedure

- Preparation of Enzyme Extract

The different samples, namely leaves, stolon and roots (0.5g), were ground with 3.0ml of potassium phosphate buffer, centrifuged at 2000g for 10 minutes and the supernatants were used for the assay.

Assay

The assay mixture contained 1.2ml of sodium pyrophosphate buffer, 0.1ml of PMS, 0.3ml of NBT, 0.2ml of the enzyme preparation and water in a total volume of 2.8ml. The reaction was initiated by the addition of 0.2ml of NADH. The mixture was incubated at 30°C for 90 seconds and arrested by the addition of 1.0ml of glacial acetic acid. The reaction mixture was then shaken with 4.0ml of n-butanol, allowed to stand for 10 minutes and centrifuged. The intensity of the chromogen in the butanol layer was measured at 560nm in a spectrophotometer (Genesys 10-S, USA).

Results

One unit of enzyme activity is defined as the amount of enzyme that gave 50% inhibition of NBT reduction in one minute.

6.1.2 Assay of Catalase (CAT)

Catalase activity was assayed following the method of Luck (1974).

Principle

The UV absorption of hydrogen peroxide can be measured at 240nm, whose absorbance decreases when degraded by the enzyme catalase. From the decrease in absorbance, the enzyme activity can be calculated.

Reagents

1. Phosphate buffer: 0.067 M (pH 7.0)
2. Hydrogen peroxide (2mM) in phosphate buffer

Procedure

- Preparation of Enzyme Extract

A 20% homogenate of the different parts of plants was prepared in phosphate buffer. The homogenate was centrifuged and the supernatant was used for the enzyme assay

Assay

H_2O_2-phosphate buffer (3.0ml) was taken in an experimental cuvette, followed by the rapid addition of 40μl of enzyme extract and mixed thoroughly. The time required for a decrease in absorbance by 0.05 units was recorded at 240nm in a spectrophotometer (Genesys 10-S, USA). The enzyme solution containing H2O2-free phosphate buffer served as control.

Results

One enzyme unit was calculated as the amount of enzyme required to decrease the absorbance at 240nm by 0.05 units.

6.1.3 Assay of Peroxidase (POD)

The method proposed by Reddy *et al.* (1995) was adopted for assaying the activity of peroxidase.

Principle

In the presence of the hydrogen donor pyrogallol or dianisidinc, peroxidase converts H_2O_2 to H_2O and O_2. The oxidation of pyrogallol or dianisidine to a colored product called purpurogalli can be followed spectrophotometrically at 430nm.

Reagents:

1. Pyrogallol: 0.05 M in 0.1M phosphate buffer (pH 6.5)
2. H_2O_2: 1% in 0.1M phosphate buffer, pH 6.5.

Procedure

Preparation of Enzyme Extract

A 20% homogenate was prepared in 0.1M phosphate buffer (pH 6.5) from the various parts of the plant, clarified by centrifugation and the supernatant was used for the assay.

Assay

To 3.0ml of pyrogallol solution, 0.1ml of the enzyme extract was added and the spectrophotometer was adjusted to read zero at 430 nm. To the test cuvette, 0.5ml of H_2O_2 was added and mixed. The change in absorbance was recorded every 30 seconds up to 3 minutes in a spectrophotometer (Genesys 10-S, USA).

Results

One unit of peroxidase is defined as the change in absorbance/minute at 430nm.

References

1. Reddy, K.P., Subhani, S.M., Khan, P.A. and Kumar, K.B. (1995). *Plant cell physiol.* **24** 987-994. Effect of light and benzyl adenine on dark-treated growing rice leaves, II changes in peroxidase activity.
2. http://shodhganga.inflibnet.ac.in/bitstream/10603/1454/13/13_chapter3.pdf
3. https://clu-in.org/download/ert/2035-R00.pdf

6.1.4 Assay of Glutathione S-Transferase (GST)

Glutathione S-transferase was assessed by the method of Habig *et al.* (1974).

Principle

The enzyme is assayed by its ability to conjugate GSH and CDNB, the extent of conjugation causing a proportionate change in the absorbance at 340nm.

Reagents

1. Glutathione (1mM)
2. 1-chloro-2, 4-dinitrobenzene (CDNB) (1mM in ethanol)
3. Phosphate buffer (0.1M, pH 6.5)

Procedure

- Preparation of Enzyme Extract

The samples (0.5g) were homogenized with 5.0ml of phosphate buffer. The homogenates were centrifuged at 5000rpm for 10 minutes and the supernatants were used for the assay.

Assay

The activity of the enzyme was determined by observing the change in absorbance at 340nm. The reaction mixture contained 0.1ml of GSH, 0.1ml of CDNB and phosphate buffer in a total volume of 2.9ml. The reaction was initiated by the addition of 0.1ml of the enzyme extract. The readings were recorded every 15 seconds at 340nm against distilled water blank for a minimum of three minutes in a spectrophotometer (Genesys 10-S, USA). The assay mixture without the extract served as the control to monitor nonspecific binding of the substrates.

Results

GST activity was calculated using the extinction co-efficient of the product formed (9.6mM−1cm−1) and was expressed as n moles of CDNB conjugated/minute.

6.1.5 Assay of Polyphenol Oxidase (PPO)

Catechol oxidase and laccase activities were estimated simultaneously by the method of Esterbauer *et al.* (1977).

Principle

Phenol oxidases are copper containing proteins that catalyse the aerobic oxidation of phenolic substrates to quinines, which are auto oxidized to dark brown pigments known as melanins. These can be estimated spectrophotometrically at 495nm.

Reagents

- Tris-HCl (50mm, pH 7.2) containing sorbitol (0.4M) and NaCl (10mM)
- Phosphate buffer (0.1M, pH 6.5)
- Catechol solution (0.01M)

Procedure

- **Preparation of Enzyme Extract**

  The enzyme extract was prepared by homogenizing 0.5g of plant tissue in 2.0ml of the extraction medium containing tris HCl, sorbitol and NaCl. The homogenate was centrifuged at 2000g for 10 minutes and the supernatant was used for the assay.

Assay

Phosphate buffer (2.5ml) and 0.3ml of catechol solution were added in the cuvette and the spectrophotometer was set at 495nm. The enzyme extract (0.2ml) was added and the change in absorbance was recorded for every 30 seconds up to 5 minutes in a spectrophotometer (Genesys 10-S, USA).

Calculations

The activity of PPO can be calculated using the formula Enzyme units in the sample = K × (DA/minute) where, K for catechol oxidase = 0.272

K for laccase = 0.242

Results

One unit of catechol oxidase or laccase is defined as the amount of enzyme that transforms 1μmole of dihydrophenol to 1μmole of quinone per minute.

6.2 Non-Enzymic Antioxidants

The non-enzymic antioxidants analyzed were ascorbic acid, α-tocopherol, total carotenoids, lycopene, reduced glutathione, total phenols, flavonoids and chlorophyll.

6.2.1 Estimation of Ascorbic Acid

Ascorbic acid was analysed by the spectrophotometric method described by Roe and Keuther (1943).

Principle

Ascorbate is converted into dehydroascorbate on treatment with activated charcoal, which reacts with 2, 4-dinitrophenyl hydrazine to form osazones. These osazones produce an orange coloured solution when dissolved in sulphuric acid, whose absorbance can be measured spectrophotometrically at 540nm.

Reagents

- TCA (4%)
- 2, 4-dinitrophenyl hydrazine reagent (2%) in 9N H_2SO_4
- Thiourea (10%)
- Sulphuric acid (85%)
- Standard ascorbic acid solution: 100μg / ml in 4% TCA

Procedure

- Extraction of Ascorbic Acid

Ascorbate was extracted from 1g of the plant sample using 4% TCA and the volume was made up to 10ml with the same. The supernatant obtained after centrifugation at 2000rpm for 10 minutes was treated with a pinch of activated charcoal, shaken vigorously using a cyclomixer and kept for 5 minutes. The charcoal particles were removed by centrifugation and aliquots were used for the estimation.

Assay

Standard ascorbate ranging between 0.2-1.0ml and 0.5ml and 1.0ml of the supernatant were taken. The volume was made up to 2.0ml with 4% TCA. DNPH reagent (0.5ml) was added to all the tubes, followed by 2 drops of 10% thiourea solution. The contents were mixed and incubated at 37°C for 3 hours resulting in the formation of osazone crystals. The crystals were dissolved in 2.5ml of 85% sulphuric acid, in cold. To the blank alone, DNPH reagent and thiourea were added after the addition of sulphuric acid. The tubes were cooled in ice and the absorbance was read at 540nm in a spectrophotometer (Genesys 10-S, USA). A standard graph was constructed using an electronic calculator set to the linear regression mode.

Results

The concentration of ascorbate in the samples were calculated and expressed in terms of mg/g of sample.

6.2.2 Estimation of Tocopherol

Tocopherol was estimated in the plant samples by the Emmerie-Engel reaction as reported by Rosenberg (1992).

Principle

The Emmerie-Engel reaction is based on the reduction of ferric to ferrous ions by tocopherols, which, with 2, 2′-dipyridyl, forms a red colour. Tocopherols and carotenes are first extracted with xylene and read at 460nm to measure carotenes. A correction is made for these after adding ferric chloride and read at 520nm.

Reagents

- Absolute alcohol
- Xylene
- 2, 2′-dipyridyl (1.2g/L in n-propanol)
- Ferric chloride solution (1.2g/L in ethanol)
- Standard solution (D, L-α-tocopherol, 10mg/L in absolute alcohol)
- Sulphuric acid (0.1N)

Procedure

- **Extraction of Tocopherol**

  The plant sample (2.5g) was homogenized in 50ml of 0.1N sulphuric acid and allowed to stand overnight. The contents of the flask were shaken vigorously and filtered through Whatman No.1 filter paper. Aliquots of the filtrate were used for the estimation.

Assay

In to 3 stoppered centrifuge tubes, 1.5ml of plant extract, 1.5ml of the standard and 1.5ml of water were pipetted out separately. To all the tubes, 1.5ml of ethanol and 1.5ml of xylene were added, mixed well and centrifuged. Xylene (1.0ml) layer was transferred into another stoppered tube. To each tube, 1.0ml of dipyridyl reagent was added and mixed well. The mixture (1.5ml) was pipetted out into a cuvette and the extinction was read at 460nm. Ferric chloride solution (0.33ml) was added to all the tubes and mixed well.

Calculations

The concentration of tocopherol in the sample was calculated using the formula,

Sample $A_{520} - A_{460}$

Tocopherols (μg) = × 0.29 × 0.15

Standard A_{520}

Results

The red color developed was read exactly after 15 minutes at 520nm in a spectrophotometer (Genesys 10-S, USA).

6.2.3. Estimation of Total Carotenoids and Lycopene

Total carotenoids and lycopene were estimated by the method described by Zakaria *et al.* (1979)

Principle

Total Carotenoids and Lycopene can be extracted in the sample using petroleum ether and estimated at 450nm and 503nm respectively.

Reagents

- Petroleum ether (40ºC - 60ºC)
- Anhydrous sodium sulphate
- Calcium carbonate
- Alcoholic potassium hydroxide (12%)

Procedure

The experiment was carried out in the dark to avoid photolysis of carotenoids once the saponification was complete. The sample (0.5g) was homogenized and saponified with 2.5ml of 12% alcoholic potassium hydroxide in a water bath at 60°C for 30 minutes. The saponified extract was transferred to a separating funnel containing 10- 15ml of petroleum ether and mixed well. The lower aqueous layer was then transferred to another separating funnel and the upper petroleum ether layer containing the carotenoids was collected. The extraction was repeated until the aqueous layer became colourless. A amount of anhydrous sodium sulphate was added to the petroleum ether extract to remove excess moisture. The final volume of the petroleum ether extract was noted. The absorbance of the yellow color was read in a spectrophotometer (Genesys 10-S, USA) at 450nm and 503nm using petroleum ether as blank.

Calculations

The amount of total carotenoids and lycopene was calculated using the formulae,

$$\text{Amount of total carotenoids} = \frac{\text{A450} \times \text{Volume of the sample} \times 100 \times 4}{\text{Weight of the sample}}$$

$$\text{Amount of lycopene} = \frac{3.12 \times \text{A503} \times \text{Volume of the sample} \times 100}{\text{Weight of the sample}}$$

Results

The total carotenoids and lycopene were expressed as mg/g of the sample.

6.2.4 Estimation of Reduced Glutathione

Reduced glutathione was determined by the method of Moron *et al.* (1979).

Principle

Reduced glutathione on reaction with DTNB (5, 5′-dithiobis nitro benzoic acid) produces a yellow colored product that absorbs at 412nm.

Reagents

- TCA (5%)
- Phosphate buffer (0.2M, pH 8.0)
- DTNB (0.6mM in 0.2M phosphate buffer)
- Standard GSH (10nmoles/ml of 5% TCA)

Procedure

- **Extraction of Glutathione:**

  A homogenate was prepared with 0.5g of the plant sample with 2.5ml of 5% TCA.

  The precipitated protein was centrifuged at 1000rpm for 10 minutes. The supernatant (0.1ml) was used for the estimation of GSH.

- **Quantification:**

  The supernatant (0.1ml) was made up to 1.0ml with 0.2M sodium phosphate buffer (pH 8.0). Standard GSH corresponding to concentrations ranging between 2 and 10 nmoles were also prepared. Two ml of freshly prepared DTNB solution was added and the intensity of the yellow color developed was measured in a spectrophotometer at 412nm after 10 minutes.

Results

The values are expressed as n moles GSH/g sample.

6.2.5 Estimation of Total Phenols

The amount of total phenols in the plant tissues was estimated by the method proposed by Mallick and Singh (1980).

Principle

Phenols react with phosphomolybdic acid in Folin-Ciocalteau reagent to produce a blue-colored complex in alkaline medium, which can be estimated spectrophotometrically at 650nm.

Reagents

- Ethanol (80%)
- Folin-Ciocalteau reagent (1N)
- Sodium Carbonate (20%)
- Standard Catechol solution (100μg/ml in water)

Procedure

The sample (0.5g) was homogenized in 10X volume of 80% ethanol. The homogenate was centrifuged at 10,000rpm for 20 minutes. The extraction was repeated with 80% ethanol. The supernatants were pooled and evaporated to dryness. The residue was then dissolved in a known volume of distilled water. Different aliquots were pipette out and the volume in each tube was made up to 3.0ml with distilled water. Folin- Ciocalteau reagent (0.5ml) was added and the tubes were placed in a boiling water bath for exactly one minute. The tubes were cooled and the absorbance was read at 650nm in a spectrophotometer (Genesys 10-S, USA) against a reagent blank. Standard catechol solutions (0.2-1ml) corresponding to 2.0-10μg concentrations were also treated as above.

Results

The concentration of phenols is expressed as mg/g tissue.

6.2.6 Estimation of Flavonoids

The method proposed by Cameron *et al.* (1943) was used to extract and estimate flavonoids.

Principle

Flavonoids react with vanillin to produce a colored product, which can be measured spectrophotometrically.

Reagents

- Vanillin reagent (1% in 70% sulphuric acid)
- Catechin standard (110μg/ml)

Procedure

- **Extraction of Flavonoids**

  The samples (0.5g) were first extracted with methanol: water mixture (2:1) and secondly with the same mixture in the ratio 1:1. The extracts were shaken well and they were allowed to stand overnight. The supernatants were pooled and the volume was measured. This supernatant was concentrated and then used for the assay.

Assay

A known volume of the extract was pipetted out and evaporated to dryness.

Vanillin reagent (4.0ml) was added and the tubes were heated in a boiling water bath for 15 minutes. Varying concentrations of the standard were also treated in the same manner. The optical density was read in a spectrophotometer (Genesys 10-S, USA) at 340nm. A standard curve was constructed and the concentration of flavonoids in each sample was calculated.

Results

The values of flavonoids were expressed as mg/g sample.

6.2.7 Estimation of Chlorophyll

The chlorophyll content in the various parts of *plants* was estimated by the method of Witham *et al.* (1971).

Principle

Chlorophyll is extracted in 80% acetone and the absorbance is measured at 645nm and 663nm. The amount of chlorophyll is calculated using the absorption coefficient.

Reagent

- Acetone (80%, pre chilled)

Procedure

Chlorophyll was extracted from 1g of the sample using 20ml of 80% acetone. The supernatant was transferred to a volumetric flask after centrifugation at 5000 rpm for 5 minutes. The extraction was repeated until the residue became colorless. The volume in the flask was made up to 100ml with 80% acetone. The absorbance of the extract was read in a spectrophotometer (Genesys 10-S, USA) at 645 and 663nm against 80% acetone blank.

Calculations

The amount of total chlorophyll in the sample was calculated using the formula.

$$\text{Total chlorophyll} = 20.2\ (A645) + 8.02\ (A663) \times \frac{V}{1000 \times W}$$

Where,

V = final volume of the extract

W = fresh weight of the leaves

Results:

The values are expressed as mg chlorophyll/g sample.

6.2.8 Evaluation of the Radical Scavenging Effects of Leaf Extracts

The scavenging effects of leaf extracts were evaluated against DPPH, ABTS, hydrogen peroxide, superoxide, nitric oxide and hydroxyl radicals.

Scavenging effects

6.2.8.1 DPPH Scavenging Effects

The ability of the leaf extracts to scavenge the DPPH radical was tested in a rapid dot-plot screening and quantified using a spectrophotometric assay.

Principle

DPPH radical reacts with an antioxidant compound that can donate hydrogen, and gets reduced. DPPH, when acted upon by an antioxidant, is converted into diphenylpicryl hydrazine. This can be identified by the conversion of purple to light yellow color.

A. Dot-Plot Rapid Assay

The rapid screening assay was performed by the method proposed by Soler-Rivas *et al.* (2000).

Reagents

- TLC plates (silica gel 60 F254-Merck)

- DPPH (0.4mM) in methanol

Procedure

Aliquots of plant extracts (3µl) were spotted carefully on TLC plates and dried for 3 minutes. The sheets bearing the dry spots were placed upside down for 10 seconds in a 0.4mM DPPH solution and the layer was dried.

Results

The stained silica layer revealed a purple background with yellow spots, which showed radical scavenging capacity.

6.2.8.2 DPPH Spectrophotometric Assay

The scavenging ability of the natural antioxidants of the leaves towards the stable free radical DPPH was measured by the method of Mensor *et al.* (2001).

Reagents

- DPPH – 2, 2-diphenyl-2-picryl hydrazyl hydrate (0.3mM in methanol)
- Methanol

Procedure

The leaf extracts (20µl) were added to 0.5ml of methanolic solution of DPPH and 0.48ml of methanol. The mixture was allowed to react at room temperature for 30 minutes. Methanol served as the blank and DPPH in methanol, without the leaf extracts, served as the positive control. After 30 minutes of incubation, the discoloration of the purple color was measured at 518 nm in a spectrophotometer (Genesys 10-S, USA).

Calculations

The radical scavenging activity was calculated as follows:

$$Scavenging\ activity\ \% = 100 - \frac{A518\ (sample)\ -\ A518\ (blank)}{A518\ (blank)} \times 100$$

6.2.8.3 ABTS Scavenging Effects

The antioxidant effect of the leaf extracts was studied using ABTS (2, 2'-azino-bis- 3-ethyl benzthiazoline-6-sulphonic acid) radical action decolonization assay according to the method of Shirwaikar *et al.* (2006).

Reagent

- ABTS Solution (7mM with 2.45mM ammonium per sulfate)

Procedure

ABTS radical cations (ABTS+) were produced by reacting ABTS solution (7mM) with 2.45mM ammonium per sulphate. The mixture was allowed to stand in the dark at room temperature for 12-16 hours before use. Aliquots (0.5ml) of the three different extracts were added to 0.3ml of ABTS solution and the final volume was made up to 1ml with ethanol.

Calculations

The absorbance was read at 745nm in a spectrophotometer (Genesys 10-S, USA) and the per cent inhibition was calculated using the formula.

$$\text{Inhibition }(\%) = \frac{(\text{Control} - \text{test}) \times 100}{\text{Control}}$$

6.2.8.4 Hydrogen Peroxide Scavenging Effects

The ability of the leaf extracts to scavenge hydrogen peroxide was assessed by the method of Ruch *et al.* (1989).

Reagents

- Phosphate buffer (0.1M, pH 7.4)
- H_2O_2 (40mM) in phosphate buffer

Procedure

A solution of H_2O_2 (40mM) was prepared in phosphate buffer. Leaf extracts at the concentration of 10mg/10μl were added to H_2O_2 solution (0.6ml) and the total volume was made up to 3ml. The absorbance of the reaction mixture was recorded at 230nm in a spectrophotometer (Genesys 10-S, USA). A blank solution containing phosphate buffer, without H_2O_2 was prepared.

Results

The extent of H_2O_2 scavenging of the plant extracts was calculated as

$$\%\text{ scavenging of hydrogen peroxide} = \frac{(A_0 - A_1) \times 100}{A_0}$$

Whereas,

A0 - Absorbance of control

A1 - Absorbance in the presence of plant extract.

6.2.8.5 Measurement of Superoxide Scavenging Activity

The superoxide scavenging ability of the extracts was assessed by the method of Winterbourn *et al.* (1975)

Principle

This assay is based on the inhibition of the production of nitroblue tetrazolium formazon of the superoxide ion by the plant extracts and is measured spectrophotometrically at 560nm.

Reagents

- EDTA (0.1M containing 1.5mg of NaCN)
- Nitroblue tetrazolium (NBT – 1.5mM)
- Riboflavin (0.12mM)
- Phosphate buffer (0.067M, pH 7.6)

Procedure

Superoxide anions were generated in samples that contained in 3.0ml, 0.02ml of the leaf extracts (20mg), 0.2ml of EDTA, 0.1ml of NBT, 0.05ml of riboflavin and

2.64ml of phosphate buffer. The control tubes were also set up where DMSO was added instead of the plant extracts. All the tubes were vortexed and the initial optical density was measured at 560nm in a spectrophotometer (Genesys, 10-S, USA). The tubes were illuminated using a fluorescent lamp for 30 minutes. The absorbance was measured again at 560nm.

Results

The difference in absorbance before and after illumination was indicative of superoxide anion scavenging activity.

6.2.8.6 Measurement of Nitric Oxide Scavenging Activity

The extent of inhibition of nitric oxide radical generation *in vitro* was followed as per the method reported by Green *et al.* (1982).

Principle

Sodium nitroprusside in aqueous solution, at physiological pH, spontaneously generates nitric oxide, which interacts with oxygen to produce nitrite ions that are estimated spectrophotometrically at 546nm.

Reagents

- Sodium nitroprusside (100mM)
- Phosphate buffered saline (pH 7.4)
- Griess reagent (1% sulphanilamide, 2% H_3PO_4 and 0.1% naphthylethylene diamine dihydrochloride)

Procedure

The reaction was initiated by adding 2.0ml of sodium nitroprusside, 0.5ml of PBS, 0.5ml of leaf extracts (50mg) and incubated at 25°C for 30 minutes. Griess reagent (0.5ml) was added and incubated for another 30 minutes. Control tubes were prepared without the extracts.

Results

The absorbance was read at 546nm against the reagent blank, in a spectrophotometer (Genesys 10-S, USA).

6.2.8.7 Measurement of Hydroxyl Radical Scavenging Activity

The extent of hydroxyl radical scavenging from Fenton reaction was quantified using 2′ deoxyribose oxidative degradation as described by Elizabeth and Rao (1990).

Principle

The principle of the assay is the quantification of 2′-deoxyribose degradation product, malondialdehyde, by its condensation with thiobarbituric acid.

Reagents

- Deoxyribose (2.8mM)
- Ferric chloride (0.1mM)
- EDTA (0.1mM)

- H_2O_2 (1mM)
- Ascorbate (0.1mM)
- KH_2PO_4-KOH buffer (20 mM, pH 7.4)
- Thiobarbituric acid (1%)

Procedure

The reaction mixture contained 0.1ml of deoxyribose, 0.1ml of $FeCl_3$, 0.1ml of EDTA, 0.1ml of H_2O_2, 0.1ml of ascorbate, 0.1ml of KH_2PO_4-KOH buffer and 20µl of plant extracts in a final volume of 1.0ml. The mixture was incubated at 37°C for 1 hour. At the end of the incubation period, 1.0 ml of TBA was added and heated at 95°C for 20 minutes to develop the colour. After cooling, the TBARS formation was measured spectrophotometrically (Genesys 10-S, USA) at 532nm against an appropriate blank. The hydroxyl radical scavenging activity was determined by comparing the absorbance of the control with that of the samples.

Results

The per cent TBARS production for positive control (H_2O_2) was fixed at 100% and the relative per cent TBARS was calculated for the extract treated groups.

References

Barros M H., B Bandy, E BTahara and A J K*owaltowski (2004). J.* Biol. Chem. 279: 49883–49888. Higher respiratory activity decreas*es mitochondria*l reactive oxygen release and increases life span in Saccharomyces cerevisiae.

Cameron G R, Milton R F and Allen J W. (1943). Lancet.179. Measurement of flavonoids in plant samples.

Elizabeth K., Rao M.N.A. (1990) Int. J. Pharm. 58: 237– 40. Oxygen radical scavenging activity of Curcumin.

Este*rbauer H., Schwarql E. a*nd Hayn M (1977) Anal. Biochem. 77, 486-494.A rapid assay for catechol oxidase and laccase using 2-nitro- 5 thiobenzoic acid.

Green L, Wagner D, Glogowski I, Skipper P, Wishnok I, Tannenbaum S. (1982) Anal Biochem 126 131-138. Analysis of nitrate, nitrite and 15Nnitrite in biological fluids

Habig W H., Pabst M J., Jakoby W B., (1974). J Biol Chem. 249: 7130-7139. Glutathione transferase: A first enzymatic step in mercapturic acid formation.

http://edblog.hkedcity.net/te_tl_e/wp-content/blogs/1685/uploads/Enriching_FST_Web_based/3b_Laboratory1_Food-related%20ExperimentForFood%20Studies_Astley.pdf

http://onlinelibrary.wiley.com/doi/10.1080/15216549700202411/pdf

http://onlinelibrary.wiley.com/doi/10.1080/15216549900202103/pdf

http://pubs.acs.org/doi/abs/10.1021/i560151a012

http://pubs.acs.org/doi/abs/10.1021/i560151a012?journalCode=iecac0

http://sydney.edu.au/science/biology/warren/docs/spec_chlorophyll.pdf

http://www.ijddr.in/drug-development/quantitative-estimation-of-phenolic-and-flavonoid-content-and-antioxidantactivity-of-various-extracts-of-different-parts-of-plumbago-zeylanica-linn.php?aid=5520

http://www.jbc.org/content/146/2/655.full.pdf

http://www.plantphysiol.org/content/plantphysiol/3/3/323.full.pdf

http://www.sciencedirect.com/science/article/pii/S2213453015000130

http://www.scientific-journals.co.uk/web_documents/3020102_peroxide.pdf

http://www.worthington-biochem.com/ty/assay.html

https://acta.mendelu.cz/media/pdf/actaun_2013061051329.pdf

https://link.springer.com/article/10.2478/s11696-007-0022-7

https://www.biovision.com/superoxide-dismutase-sod-activity-colorimetric-assay-kit.html

https://www.hindawi.com/archive/2014/918018/

https://www.hindawi.com/journals/tswj/2014/279451/

https://www.ncbi.nlm.nih.gov/pmc/articles/PMC2830880/

https://www.ncbi.nlm.nih.gov/pmc/articles/PMC2966934/

https://www.ncbi.nlm.nih.gov/pmc/articles/PMC3279904/

https://www.ncbi.nlm.nih.gov/pmc/articles/PMC3480753/

https://www.ncbi.nlm.nih.gov/pmc/articles/PMC3824142/

https://www.ncbi.nlm.nih.gov/pmc/articles/PMC549839/pdf/plntphys00428-0020.pdf

https://www.ncbi.nlm.nih.gov/pubmed/11697125

https://www.ncbi.nlm.nih.gov/pubmed/8297016

https://www.omicsonline.org/open-access/evaluation-of-free-radical-scavenging-activity-of-various-leaf-extracts-2161-1009.1000150.php?aid=33390

https://www.researchgate.net/publication/18608629_Colorimetric_Assay_of_Catalase

https://www.researchgate.net/publication/280632662_Determination_of_Brix_lycopene_bcarotene_and_total_carotenoid_content_of_processing_tomatoes_using_near_in rared_spectroscopy

Kar D., Chowdhuri D., Parmar P., R Kakkar.,. Shukla., P. K. Seth., R. C. Srima (2002).Phytotherapy Research.Volume 16,Issue 7,pages 639–645Antistress effects of bacosides ofBacopa monnieri: modulation of Hsp70 expression, superoxide dismutase and cytochrome P450 activity in rat brain.

Luck H, (1974). Catalase. In Methods of Enzymatic Analysis, Vol, II, edited by J Bergmeyer & M Grbi, Academic Press, New York. Pp 885-890.

Malik, C P., Singh, M B. (1980). Plants Enzymology and Histo-Enzymology. Kalyani Publishers, New Delhi, p. 286.

Mensor F S., Menezes G G., Leitao A S., Reis, T S., Santos C S., Coube *et al.* (2001) Phytotherapy Research, 15, pp. 127–130. Screening of Brazilian plant extracts for antioxidant activity by the use of DPPH free radical method.

Moron M S., Depierre J W., Mannervik B. (1979). ACTA. 582, pp. 67–78.Biochimica et Biophysica. Levels of glutathione, glutathione reductase and glutathioneS-transferase activities in rat lung and liver.

Roe J.H., Kuether C.A., (1943). J. Biol. Chem. 147, 399. The determination of ascorbic acid in whole blood and urine through the 2, 4-dinitrophenylhydrazine derivative of dehydroascorbic acid.

Rosenberg H R. (1992). Chemistry and physiology of the vitamins, Interscience publisher, New York. P 452-453.

Ruch R J., Cheng S.J., and Klaunig J E. (1989) Carcinogenesis, 10, pp. 1003-1008. Prevention of cytotoxicity and inhibition of intracellular communication by antioxidant catechins isolated from Chinese green tea.

Shirwaikar A, Prabhu K S, Punitha I S R. (2006) J. Exp. Biol.44:993–6In. vitroantioxidant studies ofSphaeranthus indicus (Linn)Ind.

Soler-Rivas C J. Espín C and Wichers H. J. (2000). Phytochem. Anal.11330-338. An easy and fast test to compare total free radical scavenger capacity of foodstuffs.

Winterbourne C C., Hawkins R E., Brain M and Carrel R W. (1975) J. Lab. chem. Med. 85: 337-341. The estimation of red cell superoxide dismutase activity.

Witham F H., Blaydes, D F. and Devlin, R M. (1971). Experiments in plant physiology. Litton Educational Publishing Inc. D. Van Nostrand Company. New York.

Zakaria M, Simpson K, Brown P.R, Krstulovic *(1979). A Journal* of Chromatography. 109–117. Use of reversed phase high performance liquid chromatographic analysis for the determination of provitamin A carotenes in tomatoes, 176, pp.

7

Chapter

Nucleic Acids

7.1 Extraction and Quantification Procedures of Nucleic Acids

Introduction

The search for a more efficient means of extracting DNA of both higher quality and yield as lead to the development of a variety of protocols, however the fundamentals of DNA extraction remains the same. DNA must be purified from cellular material in a manner that prevents degradation. Because of this, even crude extraction procedures can still be adopted to prepare a sufficient amount of DNA to allow for multiple end uses. DNA extraction from plant tissue can vary depending on the material used. Essentially any mechanical means of breaking down the cell wall and membranes to allow access to nuclear material, without its degradation is required. For this, usually an initial grinding stage with liquid nitrogen is employed to break down cell wall material and allow access to DNA while harmful cellular enzymes and chemicals remain inactivated. Once the tissue has been sufficiently ground, it can then be resuspended in a suitable buffer, such as CTAB. In order to purify DNA, insoluble particulates are removed through centrifugation while soluble proteins and other material are separated through mixing with chloroform and centrifugation. DNA must then be precipitated from the aqueous phase and washed thoroughly to remove contaminating salts. The purified DNA is then resuspended and stored in TE buffer or sterile distilled water. This method has been shown to give intact genomic DNA from plant tissue. To check the quality of the extracted DNA, a sample is run on an agarose gel, stained with ethidium bromide, and visualised under UV light.

Principle

The extraction of genomic DNA from plant material requires cell lysis, inactivation of cellular nucleases and separation of the desired genomic DNA from cellular debris. Ideal lysis procedure is rigorous enough to disrupt the complex starting material (plant tissue), yet gentle enough to preserve the target nucleic acid. The cetyltrimethylammonium bromide (CTAB) protocol (developed by Murray and Thompson in 1980) is appropriate for the extraction and purification of DNA from plants and plant derived foodstuff and is particularly suitable for the elimination of polysaccharides and polyphenolic compounds otherwise affecting the DNA purity and therefore quality. Plant cells can be lysed with the ionic detergent CTAB, which forms an insoluble complex with contaminants (protein,carbohydrates,lipid etc) in a high-salt environment. Under these conditions, polysaccharides, phenolic compounds and other contaminants remain in the supernatant and can be washed away. The DNA complex is solubilised by raising the salt concentration and precipitated with ethanol or isopropanol.

Major Steps in Isolation

1. Lysis of the cell
2. Isolation of DNA
3. Precipitation of DNA

1. Lysis of Cell Membrane

The first step of the DNA extraction is the rupture of the cell and nucleus wall. For this purpose, the **homogenized** sample is first treated with the extraction buffer containing EDTA, Tris/HCl and CTAB. The lysis of the membranes is accomplished by the detergent (CTAB) contained in the extraction buffer. When the cell membrane is exposed to the CTAB extraction buffer, the detergent captures the lipids and the proteins allowing the release of the genomic DNA. In a specific salt (NaCl) concentration, the detergent forms an insoluble complex with the nucleic acids. EDTA is a chelating component that among other metals binds magnesium. Magnesium is a cofactor for DNase. By binding Mg with EDTA, the activity of present DNase is decreased. Tris/HCl gives the solution a pH buffering capacity (a low or high pH damages DNA). After the cell and organelle membranes (such as those around the mitochondria and chloroplasts) have been broken apart, the purification of DNA is performed.

2. Extraction

In this step, polysaccharides, phenolic compounds, proteins and other cell lysates are removed or separated by using 24:1 ratio of chloroform isoamylalcohol. Under low salt concentration, the contaminants of the nucleic acid complex do not precipitate and can be removed by extraction of the aqueous solution with chloroform. The chloroform denatures the proteins and facilitates the separation of the aqueous and organic phases. Once the nucleic acid complex has been purified, precipitation can be accomplished.

Precipitation

In this final stage, the nucleic acid is liberated from the detergent. The aqueous solution is first treated with a precipitation solution comprising of sodium acetate, which precipitates the nucleic acid. Under these conditions, the detergent, which is more stable in alcohol than in water, can be washed out, while the nucleic acid precipitates. The successive treatment with 70% ethanol allows an additional purification, or wash, of the nucleic acid from the remaining salt.

DNA Isolation is an essential technique in molecular biology.

Isolation of High Molecular Weight (HMW) DNA is required for:

- DNA Fingerprinting
- RFLP
- Construction of genomic or sequencing libraries

- PCR Analysis
- Analysis of genome structure and gene Expression
- Quantity, Quality & Integrity of DNA directly affect the results.

Materials

Chemicals

Preparation of stock solutions of DNA extraction:

Stock Solution	Method of preparation	Function/Role
1M TrisHCl (pH 8.0) 100 ml	12.11 g Tris base was dissolved in 80 ml distilled water. The pH was adjusted to 8.0 and the total volume was adjusted to 100 ml. It was dispensed into reagent bottle and sterilized by autoclaving.	Tris/HCl gives the solution a pH buffering capacity (a low or high pH damages DNA).
0.5M EDTA (pH 8.0) 100 ml	Initially 50ml of distilled was taken in beaker and pH was adjusted above 10 by adding NaOH pellets and 18.60 g EDTA di Sodium salt was added slowly in a beaker with constant stirring by glass rod and dissolved completly. The pH was adjusted to 8.0 by adding NaOH pellets. The total volume was adjusted to 100 ml and dispensed into a reagent bottle and sterilized by autoclaving.	EDTA is a chelating component that among other metals binds magnesium. Magnesium is a cofactor for DNase. By binding Mg with EDTA, the activity of present DNase is decreased.
5M NaCl, 100 ml	29.22 g NaCl was taken in to beaker; 50 ml of distilled water dissolved well, the final volume was adjusted to 100 ml. It was taken into a reagent bottle and autoclaved.	In a specific salt (NaCl) concentration, the detergent forms an insoluble complex with the nucleic acids.
70% Ethanol, 100 ml or isopropanol	80 ml of ethanol was taken and 20 ml of distilled water was added, mixed well and dispensed into a reagent bottle and stored at 40C.	The 70% ethanol allows an additional purification, rehydration, and washing of the nucleic acid from the remaining salt. Precipitation of DNA.
Chloroform: Isoamyl alcohol (24:1), 100 ml	96 ml of chloroform (Qualigens) and 4 ml of isoamyl alcohol (Qualigens) were measured, mixed well and stored in a reagent bottle at room temperature.	The chloroform denatures the proteins and facilitates the separation of the aqueous and organic phases. Isoamyl alcohol reduces foaming & maintains the stability of the layers after centrifugation of deproteinised solution.

Ethidium Bromide (10 mg/ml), 1.0 ml	10 mg Ethidium Bromide was added to 1.0 ml of distilled water and it was kept on magnetic stirrer to ensure that the dye has dissolved completely. It was dispensed into an amber colored eppendorf tube and stored at 40C.	It binds with DNA (DNA staining) and make complex and gives fluorescence under UV to visualise DNA bands.
TBE buffer 5X (1 liter) pH 8.0	54.5 g of Tris base, 27.5 g of Boric acid (Qualigens) were taken, 20 ml of 0.5M EDTA (pH 8.0) was added. The Final volume of 1 litre was adjusted by adding distilled water and the pH was adjusted to 8.0.	Buffer not only establish a pH, but provide ions to support conductivity. If you mistakenly use water instead of buffer, there will be essentially no migration of DNA in the gel! Conversely, if you use concentrated buffer (e.g. a 10X stock solution), enough heat may be generated in the gel to melt it.

Preparation of CTAB Extraction buffer

Buffer	Method of preparation
CTAB extraction buffer (5%), 10 ml	1.0 ml of 1M TrisHCl (pH 8.0), 2.8 ml of 5M NaCl, 1 ml of 0.5 M EDTA (pH 8.0) = 0.5 gm of CTAB (w/w) , 0.2 gm of PVP (Polyvinyl pyrrolidine) (w/w) and 5.2 ml of distilled water were taken in to a flask and mixed well. The buffer was kept incubated 1t 650C for the CTAB and PVP to get dissolved. 0.15 ml (1.5%) β-mercaptoethanol (Qualigens) was added into the mixture just before use.

Plant Leaves Sample

Two types of leaves samples of Isabgol were taken for the genomic DNA extraction procedure to find out, which gives good quality and quantity DNA. One leaf sample was weighed (150mg), chopped, taken in 2ml eppendorf tube, freezed by addition of liquid N2 and kept in -70°C freezer for overnight incubation in previous day of extraction procedure. Another sample was taken as fresh in manual (Mortar and pestle) and glass rod method, but in case of tissue lyser method, in each fresh and overnight, one was taken without CTAB and one was taken with CTAB and sample IDs are as below,

Sr. No.	Sample ID	Sample details
1	A	Mortar and pestle Fresh Sample
2	B	Mortar and pestle Overnight Sample
3	C	Glass Rod Fresh Sample
4	D	Glass Rod Overnight Sample

5	E	Tissue Lyser without CTAB (Dry) Fresh
6	F	Tissue Lyser with CTAB Fresh
7	G	Tissue Lyser without CTAB (Dry) Overnight
8	H	Tissue Lyser with CTAB Overnight

Other Materials

- Microfuge tubes, 1-10 μl, 10-200 μl, 100 -1000μl (Autoclaved)
- Mortar and Pestle
- Liquid Nitrogen
- 65° C water bath
- Agarose
- 6x Loading Buffer
- Agarose gel electrophoresis system

Procedure

The first step genomic DNA extraction i.e. Lysis of cell membrane was carried out by three physical methods to find out best one.

1. Mortar and pestle
2. Glass rod and
3. Tissue lyser method.

1. Mortar and Pestle

Principle

The principle behind mortar and pestle method is that,sample is subjected to fine crushing in a powder form, but the major disadvantage this method is loss of material after grinding. Whatever quantity material we take, after grinding we are not in a condition to obtain total quantity because some material get adhered to mortar and pestle surface.

Fresh Sample

150mg of sample was weighed, chopped in small pieces and taken in mortar for fine grinding.

Liquid N_2 was added and grinding done till powder obtained by regular adding of liq. N_2.

↓

Grinded sample was taken in 2ml sterile eppendorf tube.

600μl prewarmed CTAB & 4μl of Proteinase-K added.

NOTE From CTAB addition all the steps are common to both tissue lyser and Glass rod method.

Incubated initially at 55°C for 20min and again incubated at 65°C for 60 min.

Equal volume chloroform: isoamyl alcohol (24:1) was added and mixed by gentle.

Inverting for 5-10min.

After that centrifuged at 10,000rpm for 10min at 10°C.

Supernatant was taken carefully in new eppendorf tube 150ml.

NOTE: Care was taken that, pipette tips were cut to avoid DNA shearing.

C: I wash was repeated.

Supernatant was taken into new tubes and equal volume of isopropanol was added and kept overnight in -20°C.

Next day sample was centrifuged at 10,000rpmfor 10min at 10°C supernatant was discarded and 200µl of 70% ethanolwas added.

Centrifuged at 5,000rpm for 8minat 10°C.

(Repeat this step.)

Supernatant was and pallet was air dried completely.

↓

Pallet was dissolved in 50µl TE.
Then 2µlRNAase was added and kept at 37°C for 60min.
After that kept at 65 for 60min.

Overnight Sample: B

150mg of sample was weighed,chopped and transferred into 2ml eppendorf tube.

Then 2ml of liq. N_2 was added and stored at -70°C.

Next day sample was taken in mortar and pestle from -70°C then liquid N_2 was added and grinding done till powder obtained by regular adding of liq. N_2.

↓

Grinded sample was taken in 2ml sterile eppendorf tube.

↓

➢ 600µlprewarmed CTAB &4µl of Proteinase-K added.

Next steps are same as Fresh sample.................

Glass Rod Method

Principle

The principle of glass rod method is the material is crushed in a closed system. In this closed system, nothing goes outside and nothing comes inside. In contrast to mortar and pestle method loss of material is completely avoided. In a glass rod method material is taken in a 2ml eppendorf tube then liquid N_2 is added and then crushing is done in it by a sterile glass rod. Liquid N_2 is added 2-3 times during crushing.

Protocol

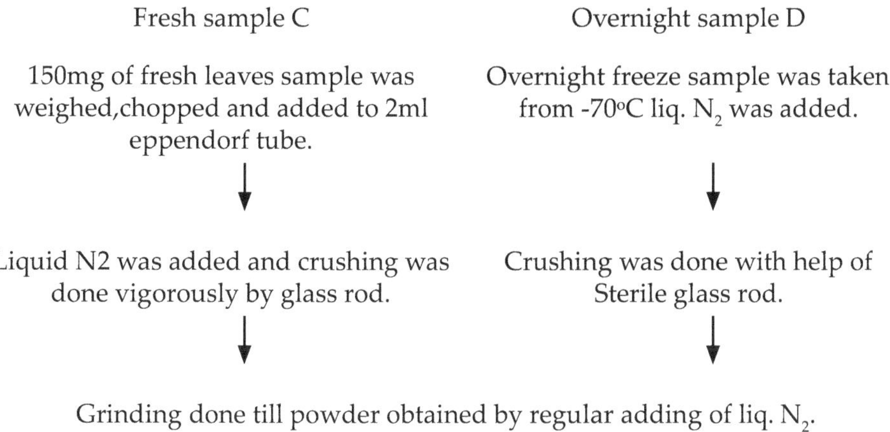

➢ 600µl prewarmed CTAB & 4µl of Proteinase-K added

Next steps are same asSample A mortar and pestle.................

3. Tissue Lyser Method

Principle

Tissue lyser induces mechanical breakage in the tissue as well as cell. It is generally used for the DNA extraction from very hard tissue e.g. Datepalm, Sorghum, etc. Tissue lyser consists of two horizontal shaking rotors having up to 30/sec frequency. In tissue lyser eppendorf tubes are allow to shake at very high frequency. In eppendorf tube two tungsten balls of, one is small sized another is larger. One ball is placed below the sample and another above the sample. During high frequency of tissue lyser shaking whatever materials (Tissues) comes in between these two balls gets crushed.

Procedure

Sample E	**Sample F**	**Sample G**	**Sample H**
Tissue Lyser without CTAB (Dry) Fresh	Tissue Lyser with CTAB Fresh	Tissue Lyser without CTAB Overnight	Tissue Lyser with CTAB Overnight
One sterile ball was Transferred into sterile sterile eppendorf tube	One sterile ball was Transferred into sterile eppendorf tube	One sterile ball was Transferred into sterile eppendorf tube	One sterile ball was Transferred into eppendorf tube
↓	↓	↓	↓
Then 150mg sample was added over it.	Then 150mg sample was added over it.	Then 150mg overnight sample was added	Then 150mg overnight sample was added
↓	↓	↓	↓
One more ball was added onto this sample	One more ball was added onto this sample	One more ball was added onto this sample	One more ball was added onto this sample
↓	↓	↓	↓
Then this tube was transferred to tissue	Then 600µl of CTAB was added	Then this tube was transferred to tissue lyser	Then 600µl of CTAB was added
	↓		↓
	Then this tube was transferred to tissue lyser.		Then this tube was transferred to tissuelyser

↓

The frequency tissue lyser was set 25/sec for one of CTAB less (**Sample E and G**) sample and 25/sec for 2min of CTAB containing samples (**Sample F and H**).

↓

After treatment 600µl CTAB was added into those sample CTAB was not added i.e. **E** and **G**

Next steps are same as **Sample A mortar and pestle............**

DNA quality confirmation through agarose gel electrophoresis

↓

1 % solution of agarose was prepared by melting 1 g of agarose in 100 ml of 1x TB buffer in a microwave for approximately 2 min.

↓

Allow to cool for a couple ofminutes then add 5µl of ethidium bromide was added, stir to mix.

↓

Cast a gel using a supplied tray and comb

↓

Allow the gel to set for a minimum of 20min at room temperature on a flat surface.

↓

Followingloaded into the into separate wells 10µL 1kb ladder

↓

5µL sample + 1µL 6x Loading Buffer

↓

Gel was run for 30 min at 100 V

↓

Gel was observed under UV light and photographs (demonstration)

↓

DNA quality and quantity was measured through Nanodrop

Results

Isolated DNA samples were run in agarose gel at 100volt for 30min and DNA bands were observed under UV light. Presence of a highly resolved high molecular weight band (Dense) indicates good quality DNA, presence of a smeared band indicates DNA degradation.

Sample A (Mortar and pestle Fresh Sample)

It was fresh sample and DNA extracted by using mortar and pestle. Band obtain of Sample is very light which indicates that low quantity of DNA is present but quality is good. Material was remained attached tothe mortar and pestle surface due to which DNA was obtained in low quantity.

Sample B (Mortar and pestle Overnight Sample)

The band obtained by sample B which was kept -70°C in liquid N2 indicates RNA contamination. The reason behind RNA contamination was the wing occurred due to loss of power during overnight incubation period in -70°C freezer.

Sample C (Glass Rod Fresh Sample)

The band obtained from this sample is dense and clear and no smear was observed, presence of a highly resolved high molecular weight band (Dense) indicates good quality DNA. The band is dense is due to crushing in closed system in contrast to mortar and pestle.

Sample D (Glass Rod Overnight Sample)

That was overnight incubated in -70°C. The band obtained is light and reason is same as sample B i.e. thawing.

Sample E (Tissue Lyser without CTAB (Dry) Fresh)

Smeared band is observed, which indicates shearing of DNA and protein contamination.

Sample F (Tissue Lyser with CTAB Fresh)

The band obtained from sample F is very light is due to thawing during incubation period.

Sample H (Tissue Lyser with CTAB Overnight)

Smeared band was observed, which is due to shearing of DNA. It was also a thawed sample.

7.2 DNA Quantity and Quality Measured by Nanodrop

All the samples were subjected quantity and quality measurement by nanodrop, which takes two different absorbance at a time, i.e. 260/280 and 260/230.

➢ 260/280

- The ratio of absorbance at 260 nm and 280 nm is used to assess the purity of DNA andRNA. A ratio of ~ 1.8 is generally accepted as "pure" for DNA; a ratio of ~ 2.0 is generally accepted as "pure" for RNA.
- If the ratio is appreciably lower in either case, it may indicate the presence of protein, phenol or other contaminants that absorb strongly at or near 280 nm.

➢ 260/230

- This ratio is used as a secondary measure of nucleic acid purity. The 260/230 values for "pure" nucleic acid are often higher than the respective 260/280 values. Expected 260/230 values are commonly in the range of 2.0-2.2.
- If the ratio is appreciably lower than expected, it may indicate the presence of contaminants which absorb at 230 nm.

Nanodrop Readings

Sample ID	260/280	260/230	Quantity(ng/μl)
A	1.82	1.99	68.4
B	2.06	1.63	180.7
C	1.92	2.18	326.1
D	1.89	1.34	125.9
E	1.62	0.65	80.6
F	1.80	1.27	132.4
G	1.65	0.58	24.1
H	1.89	1.54	378.8

Sample A-

Reading at both 260/280 and 260/230 is between 1.8 - 2.0 which indicates good quality/pure DNA.

Sample B-

At 260/280 are 2.06 which is acceptable and slight RNA contamination but at 260/230 is 1.63 which indicates protein contamination. Quantity is 180.7ng/μl.

Sample C-

Reading at 260/280 is 1.92 and at 260/230 is 2.18 which shows pure DNA. Quantity is 326.1ng/μl, which is also good. In agarose gel electrophoresis a dense band was observed which indicates high molecular weight DNA in good quantity.

Sample D-

Reading at 260/280 is good but at 260/230 is 1.34 which indicates protein contamination. Quantity is not good which is also indicated in AGE by a light band.

Sample E-

Reading at both 260/280 and 260/230 is below 1.8 which indicates more protein contamination quantity is also poor.

Sample F-

Reading at both 260/280 is 1.80 which is good but at 260/230 below 1.8 which shows protein contamination. Quantity is also not good.

Sample G-

Reading at both 260/280 and 260/230 is below 1.8 which indicates more protein contamination quantity is also poor as indicated in AGE very light band was observed(negligible).

Sample H-

Reading at 260/280 is good but at 260/230 is 1.34 which indicates protein contamination.

Conclusion

Among all the results obtained by agarose gel electrophoresis and nanodrop sample C (Glass Rod Fresh Sample) is best as indicated in AGE by dense clear band and also in nanodrop reading, at 260/280 is 1.92 and at 260/230 is 2.18 which shows pure DNA. Quantity is 326.1ng/µl, which is also good In case of tissue lyser, quantity as well as quality was low. Reason behind this that tissue lyser conditions are not standardised. To obtain good quantity and quality DNA tissue lyser condition are to be standardised by testing sample at different frequencies/ sec. for different period of time.

7.3.1 DNA Quantification Through NanoDrop

Introduction

Thermo Scientific NanoDrop 1000 spectrophotometer and NanoDrop 8000 spectrophotometer will include the absorbance of all molecules in the sample that absorb at the wave length of interest. Since nucleotides, RNA and DNA absorb at 260 nm, they will contribute to the total absorbance of the sample. Therefore, to ensure accurate results, nucleic acid samples will require purification prior to measurement.

Principle

The Beer-Lambert law:

It states that the absorbance is directly proportional to the path length, b of the sample and its concentration, c.

$A = abc$

Ratios to Measure Nucleic Acid Purity

1. 260/280

- The ratio of absorbance at 260 nm and 280 nm is used to assess the purity of DNA and RNA. A ratio of ~1.8 is generally accepted as "pure" for DNA; a ratio of ~2.0 is generally accepted as "pure" for RNA.
- If the ratio is appreciably lower in either case, it may indicate the presence of protein, phenol or other contaminants that absorb strongly at or near 280 nm.

2. 260/230

- This ratio is used as a secondary measure of nucleic acid purity. The 260/230 values for "pure" nucleic acid are often higher than the respective 260/280 values. Expected 260/230 values are commonly in the range of 2.0-2.2.
- If the ratio is appreciably lower than expected, it may indicate the presence of contaminants (phenolic or protein compounds) which absorb at 230 nm.

Procedure

1. Initialize the instrument
2. Set the blank

3. 1 μ of the sample
4. Click on measure

Observations

Sr. No.	260/280	260/230	ng/μl
1			
2			
3			
4			
5			

7.3 Extraction of RNA byTrizol Method

Introduction

RNA in expansion is a ribonucleic acid which is a biological macromolecule that serves in number of different functions. A messenger RNA transcribed from DNA serves as template for synthesis for protein. Protein synthesis is carried out by ribosomes which consist of ribosomal RNA and proteins. Amino acids for protein synthesis are delivered to ribosomes machinery on transfer RNA molecules RNAs are also part of ribo-protiens involved in RNA processing. A typical mammalian cell contains ~10-5 μg of RNA, 80-85% of which is ribosomal RNA(chiefly28s,18s,5.8s&5s)while 15-20% consists of a variety of low molecular weight species(e.g. transfer RNAs and small nuclear RNAs). These abundant RNAs are of defined size and sequence and can be isolated in virtually pure form by gel electrophoresis, density gradient centrifugation, anion exchange chromatography or HPLC. In total RNA 1 to 5% is m-RNA. Most eukaryotic m-RNAs carry at their 3′ termini a tract of polyadenylic acid residues.

Principle

For extraction of RNA, lysis of tissue and disruption of cell wall is most important thing. For RNA isolation 3 contaminations which occur during RNA isolation.

1. Protein
2. RNase
3. DNA

As ribose residues carry hydroxyl groups in both the 2′ and 3′ positions, RNA is chemically much more reactive than DNA and is easily cleaved by contaminating RNases enzymes with various specificities that share the property of hydrolyzing diester bonds linking phosphate and ribose residues. RNase do not require any cation for activity and thus cannot be easily inactivated by the inclusion of EDTA or other metal ions chelators in buffer solutions and many RNases cannot get denature at high temprature. Many methods use strong denaturants such as guanidinium hydrochloride or guanidinium thiocyanate to disrupt cells, solubilize their components and denature endogenous RNase simultaneously. Diethylpyrocarbonate(DEPC), a highly reactive alkylating agent, is used to inactive RNases in buffers on glassware. Phase separation is done using chloroform/phenol and RNA remain in upper aqueous phase.

Quantification Principle

Concentration of RNA is determined by measuring the absorbance at 260nm corresponding to 40µg of RNA per ml.

A_{260}=1=40 µg/ml

This relation is rated only for measurements made at netural pH.

Reaction Component

- DEPC (diethylpyrocarbonate)
- Trizol reagent
- Chloroform
- Ethanol
- Liquid nitrogen

Protocol

1. Take 80-100 mg of sample in chilled morter and pestle (on ice box)
2. Crush the sample for 20 mins with liquid nitrogen.
3. Add 900-1000 µL of trizol reagent and crush the sample for 5 mins on ice.
4. Take the brownish coloured liquid in 2 ml eppendorf tube and incubate it on ice for 5 mins.
5. Add 200 µl of chloroform & shake vigorously and incubate for 3-4 mins.
6. Centrifuge it at 16000 rpm for 15 mins at 2-8 ºC
7. Transfer aqueous upper phase in to new tube and add equal volume of chilled isopropanol and incubate it on ice for 10 mins.
8. Centrifuge it at 10600 rpm for 10 mins at 2-8 ºC
9. Remove supernatant and wash the pellet with 75% ethanol (vortex briefly)
10. Centrifuge it at 8000-9000 rpm for 8 mins at 2-8 ºC
11. Decant the supernatant and air dry the pellet for 5-10 mins(do not over dry)
12. Dissolve the pellet in nuclease free water (DEPC treated water)

Result and Discussion

Extraction of total RNA requires quantification that has to be measure using nanodrop. After doing quantification ratio of 260/280 and 260/230 for total RNA is describe below.

Sr. No.	Sample	260/280	260/230	ng/µl
1.	Rice	1.95	1.67	233.6
2.	Rice	1.79	0.51	172.0

Conclusion

Extraction of RNA using trizol method gives total RNA that are ribosomal RNA, messenger RNA, and transfer RNA. DEPC is used to inhibit RNase in buffer and guanidinium thiocyanate is used to disrupt cells, solubilize their components, and denature endogenous RNase simultaneously. Using chloroform separation of phase is done. RNA remains in upper aqueous phase. That RNAs having their own half life varies in hours to days. In that m-RNA having half life in hours and it is more important than other RNAs. So purification of mRNA is the further most important step after extraction of total RNA and that is used to make cDNA libraries.

References

Ausubel F. M. R. Brent, R. E. Kingston, D. D. Moore, J.G. *Seidman, J. A. Smith,* K. Struhl (eds.) (2003). Current Protocols in Molecular Biology. John Wiley & Sons,Inc.

Brown T.A. (1998). Molecular Biology LabFax. Gene Analysis. Academic Press.

Edwards D. (2007). Plant Bioinformatics. Methods and Protocols Humana Press, New Jersey.

Farrell R. E. (2005). RNA methodologies. ALaboratory Guide for Isolation and Characterization Elsevier Academic Press.USA.

himedialabs.com/TD/HTBM005.*pdf*

*http://dx.doi.org/10.5402/2012/*205049

O'Connell J. (2002). RT- PCR Protocols Humana. Press, New Jersey.

Sambrook. *J. and D. W.* Russell (2001). Molecular Cloning. A Laboratory Manual. CSHL *Press, New York.*

*Su*rzycki S. (2000). Basic Techniques in Molecular Biology, Springer, *Germany.*

Weising K., H. Nybom, K. Wolff and G. Kahl. (2005). DNA fingerprinting in plants. Principles, Methods, and Applications. CRC Press, Taylor & Francis Group

William Wu (2004). Gene Biotechnology. CRC Press, USA.

8

Chapter

Extraction and Quantifications of Pigments

8.1 Determination of Total Chlorophyll

The chlorophyll is the essential components for photosynthesis and occurs in chlorophyll as a green pigments in all photosynthetic plant tissues. Chemically, each chlorophyll molecules contains a porphyrin (tetra pyrol) nucleus with the chelated magnesium atom at the centre and a long-chain hydrocarbon (phytyl) side chain attached through a carboxylic acid group. There are at least five types of chlorophyll in plants.

The methodologies used for chlorophyll extraction in plant materials are almost always based on methods that destructively extract leaf tissue using organic solvents that include acetone, dimethylsulfoxide (DMSO), methanol,*N*, *N*-dimethyl formamide and petroleum ether, During the extraction and dilution, significant pigment losses may occur thus leading to a high variability in the results.

Extermination of Chlorophyll Contents Form Leaf by Acetone Methods

Total chlorophyll, chlorophyll a and b contents were determined by the methods of Arnon (1949).

Principle

Chlorophylls are extracted in 80% acetone and the absorption at 663nm is read in a spectrophotometer. Using the absorption coefficients, the amount of chlorophyll is calculated.

Materials

- 80% Acetone

Procedure

1. Weigh 1 g of finely cut and well mixed representative sample of leaf or fruit tissue into a clean mortar.
2. Grind the tissue to a fine pulp with the addition of 200ml of 80% acetone.
3. Centrifuge at 5000 rpm for 5 minutes and transfer the supernatant to a 10 ml volumetric flask.
4. Grind the residue with 20 ml of 80% acetone, centrifuge and transfer the supernatant to the same volumetric flask.

5. Repeat this procedure until the residue is colorless.
6. Wash the mortal and pestle thoroughly with 80% acetone and collect the clear washing in the volumetric flask.
7. Make up the volume to 100ml with 80% acetone.
8. Read the absorbance of the solution at 645, 663 and 652nm against the solvent ()80% acetone) blank.

Calculation

Calculate the amount of chlorophyll present in the extract mg chlorophyll per g tissue using the following equations:

$$\text{Mg of chlorophyll a /g tissue} = 12.7\,(A663) - 2.69\,(A645) \times \frac{V}{1000 \times W}$$

$$\text{Mg of chlorophyll b/g tissue} = 22.9\,(A645) - 4.68\,(A663) \times \frac{V}{1000 \times W}$$

$$\text{Mg of total chlorophyll/ g tissue} = 20.2\,(A645) + 8.02\,(A663) \times \frac{V}{1000 \times W}$$

Notes

The amount of tissue taken for extraction may be varied. Accordingly amount of 80% acetone used may be alter so that the final extract has a volume based on 10mg plant materials extracted in 1ml of acetone.

Extermination of Chlorophyll Contents Form Leaf by DMSO Methods

Several methods are commonly used to extract chlorophyll (Chl) from plants. There are many reasons why particular extraction methods are used, but this inherent methodological variation may create several problems in making comparisons between species or different studies, potential biases in chlorophyll extraction may be due to solvent types, solvent purity, tissue type and degree of leaf maceration and the equation used to calculate the chlorophyll concentrations.

Principle

Dimethylsulfoxide (DMSO) appears to be a reliable solvent for extracting chlorophyll (Chl), however, modification of standard methods may be necessary for some species under field conditions. We found that Chlorophyll extraction of whole leaf tissue with DMSO incubated at between 25 and 40 °C was generally similar to the 80% acetone method, except for one graminoid species that required maceration. There was little effect of incubation temperature or duration of incubation beyond 7 h on extraction efficiency, but DMSO extracts were less stable than acetone extracts during one week of cold storage, especially if they thawed during this period. Since Chlorophyll extraction methods may provide variable results, particularly in the field.

Materials

- Di-methyl-sulfoxide
- Fully expanded leaf

Procedure

1. Weigh the leaf tissue (20mg) from each leaf was cut into small (40-6mm2) species.
2. Take theses pieces into a vial with 7 ml of DMSO.
3. Make the five replication of each sample
4. Incubate the sample containing tubes at 40C for 1 hour.
5. Add 3ml of DMSO in sample (Final Volume must be 10 ml)
6. Centrifuge the sample at 1000 rpm for 10 minutes.
7. Avoid the direct expose of sunlight
8. Read the absorbance of the solution at 645, 663 and 652nm against the solvent.

Calculations

Chlorophyll concentrations were calculated using Wellburn's (1994) equation for a low resolution spectrophotometer.

DMSO

$$Chl_a = 12.19A_{665} - 12.19\ A_{649}$$

and

$$Chl_b = 21.99A_{663} - 5.32A_{665}.$$

8.2.1 Estimation of Anthocynine From Plant Tissues

This method describes the measurement of total anthocyanins in plant tissue based on the methods described by Iland *et al.* (1996, 2000). It involves extraction of these compounds from a homogenised plant sample, expression of their color at low pH and quantification based on their absorbance in the visible region of the light spectrum. Malvidin-3-glucoside is the major anthocyanin in Vitis vinifera grapes but is not the only anthocyanin and the results are expressed in malvidin-equivalents for comparative purposes only. A minimum sample size of 50 g (approximately representative 50 berries) is required for this determination.

Materials and Methods

1. Calibrated spectrophotometer (traditional, plate reader, or other type) capable of measuring absorbance at 520 nm with an accuracy of ±2 nm. Note that if using a style other than traditional, the calculation may differ from that presented here and must be validated before use.
2. Cuvettes suitable for the instrument being used, for example, 10 mm path length quartz, optical glass or acrylic cuvettes (4 mL capacity). For large sample numbers, disposable acrylic cuvettes offer convenience, as the final dilutions can be made directly into the cuvettes.
3. Liquid handling devices for accurate delivery of 200 μL, 3.8 mL and 10 mL volumes
4. Analytical balance with minimum scale reading of 0.001 g.

5. Homogeniser appropriate to the sample size:
6. For small sample sizes (50-100 g) an Ultra- Turrax T25 high-speed homogeniser with an S25N dispersing head can be used.
7. Centrifuge capable of a radial centrifugal force(RCF) of at least 1800g (i.e. 4,000 rpm with a 10cm rotating radius).
8. Mixing device such as a rotary suspension mixer,shaker table or roller mixer.

Reagents

1. 1.0 M Hydrochloric acid.
2. 50% v/v ethanol in Milli-Q (or equivalent) water.

Procedure

1. Samples can be analysed fresh or can be frozen (-20°C) prior to analysis. Fresh samples must be stored cool (approximately 4°C) and analysed within 24 hours of collection. For field samples, this can be achieved, for example, by storing freshly collected samples in cooler boxes with ice packs until delivery to the laboratory. Short term freezing of whole sample (i.e. overnight) does not significantly affect the color of sample, although there may be some loss of color with sample that have an anthocyanin concentration of greater than 1.9 mg/g. There is no significant color loss during storage of frozen whole sample for at least 3 months. On the other hand, homogenates of red sample lose color (approximately 0.1 mg/g) after even overnight freezing.
2. If the sample is large, a representative sub-sample must be taken. This is best done with frozen grapes to avoid juice loss, using the following procedure.
   - If the sample contains bunches, remove all berries from the rachis by hand and place into a tray or container. If the berries are loose, just place all berries into a tray or container;
   - Gently mix the berries by hand;
   - Randomly select berries from different areas within the container to make up the required sample size (e.g. 200 berries, or approximately 200 g). Further replicates can be taken if required and stored frozen for reference.
3. Usually the determination is reported on a mass basis (i.e. per gram of grapes), but if the determination is required to be expressed on a 'per berry' basis, a sample of a defined number of berries (e.g. 50- berry sample) is weighed to determine the average berry weight. Record the berry weight to at least two decimal places or as appropriate to the balance being used.
4. If frozen, berries are thawed (usually overnight in a refrigerator at 4°C) and processed cold (less than 10°C) to minimize oxidation of color components.
5. Berry samples are homogenised according to the following settings:

- **Ultra-Turrax:** Process at the maximum speed setting (24,000 rpm) for approximately 30 seconds, scrape the homogenate from the homogeniser shaft into the vessel, and then homogenise for a further 30 seconds. Ensure that all seeds are thoroughly macerated and all homogenate is scraped from the homogeniser shaft and collected in the homogenising vessel.
- **Retsch:**

  Process the sample at a speed of 8,000 rpm for 20 seconds.
- **Waring:**

  Process for 60 seconds on high speed

  Comparative trials with these three blenders have shown no significant differences with total anthocyanin measurement, but the Waring blender did show reduced recovery of total Phenolic, possibly due to the fact that it does not homogenise grape seeds as effectively as the other types. Please note that the homogenisation step can give rise to a significant safety hazard. Please ensure that adequate attention to safety and all recommendations of safety precautions and maintenance provided by the manufacturer is considered when using these types of equipment.

6. Once homogenised, samples must proceed to the extraction step within 4 hours.
7. Mix the homogenate well, and then transfer approximately 1 g to a pretared plastic 10 mL centrifuge tube. Record the weight (to at least two decimal places or as appropriate to the balance being used) of the homogenate portion for use in calculations.
8. Add 10 mL of 50% v/v aqueous ethanol to the portion of homogenate, cap the tube and agitate on a mixing device for one hour. Alternatively, mix by inverting the tube regularly (approximately every 10 minutes) over a period of one hour. Ensure that mixing is efficient and that the pellet does not become lodged in the bottom of the tube.
9. Centrifuge the homogenate/ethanol mixture at 1800g for 10 minutes. The supernatant is now termed the "extract". (This same extract can also be used for the determination of total Phenolic.)
10. If approximately 1 g of homogenate was used (i.e. 0.95 to 1.05 g), the final extract volume is estimated as 10.5 mL. Alternatively, this volume can be measured with a suitable volume measuring container.
11. The extract can be stored in the freezer (at -20°C) for up to 3 months without significant loss of color.
12. Transfer 200 μL of extract to an acrylic cuvette (10 mm path length), add 3.8 mL of 1.0 M HCl, cover with Para film then mix by inverting. Incubate at room temperature for at least 3 hours, but not longer than 24 hours. This step is critical in allowing full expression of color.

13. The incubation time can be reduced if the 50% ethanol solution (i.e. that used in step 9) is adjusted to pH 2.0 with HCl, however if the extract is also required for a glycosyl-glucose assay, the 50% ethanol solution must not be acidified. Note that some industry practitioners have reduced the incubation time using other alternative procedures, but these must be validated before use. It may also be necessary to adjust the dilution to ensure that the absorbance is within the working range of the spectrophotometer, and any such change to the procedure must be validated before use.
14. Measure absorbance of the acidified diluted extract at 520 nm using a 1.0 M HCl blank.

Calculations

$$\text{Anthocyanins}\ (\text{mg/g}) = \frac{A_{520} \times \text{DF x final extract volume}\ (\text{mL}) \times 1000}{500 \times 100 \times \text{homogenate weight}\ (\text{g})}$$

Whereas;

1. absorbance at 520 nm, 10 mm path length (Procedure step 14)
2. Dilution factor (DF) at procedure step 13 (dilution of extract in 1 M HCl): 20 in this case.
3. if the weight of homogenate sample taken at procedure step 8 is kept within a 5% range (0.95 – 1.05 g), a standard figure of 10.5 mL can be used as the final extract volume.
4. Absorbance of a 1% w/v (1 g/100 mL) solution of malvidin-3- glucoside, 10 mm path length (Somers and Evans, 1974) – this is a value traditionally used within the Australian wine industry. Using a molecular weight of 529 for malvidin-3-glucoside, equates to a molar absorbance of 26,455. Literature values for this vary due to many confounding factors such as non-linearity resulting from co pigmentation at high concentrations, temperature, pH effects and the difficulty in obtaining stable reference material of high purity. Malvidin-3-glucoside is the major anthocyanin in Vitis vinifera grapes but is not the only anthocyanin and the results are expressed in malvidin-equivalents for comparative purposes only.

8.2.2 Estimation of Anthocyanins by Using UV-Visible Spectroscopy by Using Ph- Differential Methods

Principle

Anthocyanin pigments undergo reversible structural transformation with changes in pH manifested by strickigly different absorbance spectra. The colored oxonum form predominanted at ph 1 and colored hemi-ketal form at pH 5. The pH differential form or method is based on this reaction.

Materials

1. **0.025 potassium chloride buffer (pH1):**

   Mix 1.86g of KCl and 980 ml of distilled water in a beaker. Measure the pH and adjust at pH 1 with the concentrated HCl; transfer to 1 liter volumetric flask and final make up the volume up to 1 liter.

2. Mix 54.43 g of CH_3COONA in 960 ml of distilled water and adjust to the pH 4.5 with concentrated HCl and transfer to 1 liter volumetric flask and make up the volume to 1 liter.

Procedure

1. For Sample preparation please see the above described method.
2. Take two test tubes and add the 0.3 ml sample (diluted if required) in each tubes.
3. Make 3 replication of each sample
4. Add KCl of pH1 (4.7 ml) to one tube
5. Add CH_3COONA of pH 4.5 (4.7 ml) to second tube
6. All these addition should be down in a dark room
7. Equilibrate for 15 minutes
8. Take the absorbance at 520 and 700 nm between 15 minutes to 1 hour after sample preparation in a dark.
9. Absorbance readings are made against water blank.

Calculation

1. Calculate absorbance of diluted sample (A) as:

   $A = (A_{520} - A_{700})_{PH\text{-}1} - (A_{520} - A_{700})_{PH\ 4.5}$

2. 2. Calculate monomeric anthocyanins pigment concentration in original sample

$$\text{Monomeric anthocyanins} = \frac{A \times Mw \times DF \times 100}{E \times 1}$$

Whereas;

Mw = Molecular weight

Df = Dilution Factor

Mw and E used in the formula corresponds to the predominant anthocyanins in the sample i.e. Marridin -3- glucoside → Mw = 493.5; E = 28000

Or Cyanidin- 3- glucoside → Mw = 449.2; E = 26900

8.3 Estimation of Carotenoids

Carotenoids, the tetraterpenoid (C_{40}) compounds, are unbiquitous in plants. Carotenoids are synthesized exclusively by photosynthetic organisms including crop plants, algae, few fungi and certain bacteria where they play a vital role in plant metabolism and bio-synthesis of other bio-molecules. Aside from providing aesthetic qualities as colorants in the plant and the animal kingdoms, these pigments also play significant role in photo system-I and photo system-II. Most importantly, they acts as light harvesters and protect chlorophylls from photo oxidation.

Principle

The total carotenoids are extracted and partitioned in organic solvents on the basis of their solubility. The separation of individual componants is effected by chromatography on activated magnesia.

Materials

- Acetone-dry, alcohol-free. To dry treat with anhydrous sodium sulphate and distil over granular zinc (10 mesh).
- Hexane- distilled over KOH
- Activated Magnesia
- Diatomaceous Earth – Hyflo Supercel
- MgC03

Procedure

1. Finely cut materials with scissors/knif or grind in food chopper to assure representative sample. (if analysis can not be performed immediately, blanch in boiling water for 5- 10 minutes and stired in frozen)
2. Place 2- 5g weighed sample in a high- speed blender. Add 40ml acetone, 60ml hexane and 0.1g $MgCO_3$. Blender for 5 minutes.
3. Filter with suction or let residue settle and decant into a separator.
4. Wash the residue twice with 25ml portions acetone, then with 25ml hexane and combine with extract.
5. Wash acetone from extract with five 100ml portions H_2O, transfer upper layer to 100ml volumetric flask containing 9 ml acetone, and dilute to volume with hexane.
6. Pack activated magnesia-diatomaceous earth (1:1) mixture in a chromatographic tube (17.5 x 2.2 cm OD). To prepare the column, place small glass wool or cotton plug inside the tube, add loose adsorbent to 15cm depth, attach tube to suction flask and apply full vacuum of water pump. The packed column should be about 10cm high. Place 1cm layer of anhydrous sodium sulphate above adsorbent.
7. Pour the extract into the column with vacuum continuously applied to the flask.
8. Use 50ml of acetone-hexane (1:9) mixture to develop the chromatogram and wash visible carotenes through adsorbent. Keep the top of the column covered with solvent durong entire operation.
9. Collect the entire elute. (carotenes pass rapidly through the column; band of xanthophylls, carotene oxidation products and chlorophylls are retained.)
10. Transfer the elute, which has been reduced in volume by the loss of vapour through water-pump, to a 100ml volumetric flask, dilute to volume with the acetone-hexane mixture.

11. Determine the absorbance of the solution as soon as possible with a spectrophotometer at 436nm. Calibrate the instrument first with solutions of different concentrations of high purity β-carotene.
12. Prepare the calibration curve and calculate the carotene content (mg/100g) in the sample material using the curve.

Notes

1. Alcohol may be used instead of acetone for extraction. Use 80ml of alcohol and 60ml of hexane in blender; other volumes are same as for acetone extraction.
2. The extract volume will get reduced due to evaporation at room temperature. Maintain the volume at appropriate stages.
3. Carotenoids that are bound as ester are not estimated in this procedure. A portion of the extract may be hydrolyzed using methanolic KOH in a water bath for 5-10 minutes and then subject to chromatography.

8.4 Estimation of Lycopene

Lycopene is responsible for the red color of tomato and freshly part of the water melon. It is a carotene having the formula C40H56.though it has no nutritional value, its contribution to the color of tomato has great role in consumer acceptability.

Principle

The carotenoids in the sample are extracted in acetone and then taken up in petroleum ether. Lycopene has absorption maxima at 473nm and 503nm. One mole of lycipene when dissolved in one liter petroleum (40 to 60°C) and measured in a spectrophotometer at 503 nm in 1cm light path gives an absorbamce of 17.2 x 10^4. therefore, a concentration a of 3. 1206 μg lycopene/ml gives unit absorbance.

Materials

- Acetone
- Petroleum Ether 40-60
- Anhydrous Sodium Sulphate
- 5% Sodium Sulphate

Procedure

1. Take 3-4 tomato fruits (sample) and pulp it well to a smooth consistency in a waring blender.
2. Weigh 5 -10 g of this pulp.
3. Extract the pulp repeatedly with acetone using pestle and mortar or a waring blender until the residue is colorless.
4. Pool the acetone extracts and transfer to a separating funnel containing about 20ml petroleum ether and mix gently.
5. Add about 20 ml of 5 % sodium sulphate solution and shake the separating funnel gently. Volume of petroleum ether might be reduced during these processes because of its evaporation. So add 20ml more of petroleum ether

to the separating funnel for clear separation of two layers. More of the color will be noticed in the upper petroleum ether layer.

6. Separate the two phases and re-extract the lower aqueous phase with additional 20ml petroleum ether until the aqueous phase is colorless.
7. Pool the petroleum ether extracts and wash once with a little distilled water.
8. Pour the washed petroleum ether extract containing carotenoids into a brown bottle containing about 10 g anhydrous sodium sulphates. Keep it aside volumetric flak.
9. Decant the petroleum ether extract into a 100ml volumetric flask through a funnel containing cotton wool. Wash sodium sulphate slurry with petroleum ether until it is colorless and transfer the washings to the volumetric flask.
10. Make up the volume and measure the absorbance in a spectrophotometer at 503nm using petroleum ether as blank.

Calculation

$$\text{Mg lycopene in 100g sample} = \frac{31.206 \times \text{Absorbance}}{\text{Wt. of sample }(g)}$$

8.4 Estimation of Curcumin

Curcumin is very important due to its use in foods, cosmetics and medicines. It is added to foods such as fats; impart color, meat products and beverages, to impart color. It is also used as an antioxidant to prevent the rancidity (Stankovic 2004). It is used as medicine to treat digestive disorders, purify the blood, kill germs, protect liver and reduce cholesterol level, anti HIV, anti-cancer, anti-thrombosis and anti-Alzheimer (Khanna 1999). Curcumin has different functional groups such as parahydroxy, keto and double bonds, which are responsible for its antioxidative, anti-inflammatory, anti-cancer and anti-mutagen activities.

Principal

Curcumin is quantitatively extracted by refluxing the material in alcohol and is estimated spectrometrically at 425nm.

Materials

Extraction of Curcumin:

Turmeric (*Curcuma longa* L.) powder used for the extraction of curcumin was procured from the local market. Turmeric powder was passed through different sieves (0.42, 0.50, 0.60, 0.71 and 0.85mm) to get desired particle size.

Procedure

Take turmeric powder 1 g from each particle size and a measured amount of absolute alcohol were transferred to double jacketed flasks (length 18.5 cm, diameter 6.5cm), which was simultaneously stirred and heated using a water bath at a selected temperature and predetermined time. Absorbance of sample was taken at 425 nm using UV-Spectrophotometer and amount of curcumin was calculated from standard curve.

Preparation of Standard Curve

A Stock solution was prepared by dissolving 10 mg of curcumin in absolute alcohol to get concentrated of 1 mg/ml. different concentrations (0.001 to 0.005 mg/ml) were made by diluting the stock solution with absolute alcohol. The absorbance was read at 425 nm and plotted against concentration.

Calculations

$$\text{Curcumin }\% = \frac{\text{Curcumin extracted }(g) \times 100}{\text{Turmeric used }(g)}$$

References

Alpert P (1984) Bryologist87: 363-365. Analysis of chlorophyll content in mosses though extraction in DMSO.

Arnon D I (1949) Plant physiology. 24 1.

Bauernfeind J. C. 1972. J. Agric. Food Chem. 20: 456-473.Carotenoid vitamin A precursors and analogs in foods and feeds.

Bendich, A. and G.S. Higdon. 2004. Am. J. Nut. 32(2):225-230.Biological actions of Beta-Carotene.

Britton, G. 1995. Amer. J. Nut. 47(2):1990-2001. Carotenes by their chemical structure.

Durst R W and Wrolstad R. E. (2005). Unit F1.2: Characterization and measurement of anthocyanins by UV–visible spectroscopy. In R. E. Wrolstad (Ed.), Handbook of analytical food chemistry (pp. 33–45). New York: John Wiley & Sons.

Francis F. (1982). Analysis of anthocyanins. In P. Markakis (Ed.) Anthocyanins as food colors. New York: Academic Press.

Giusti M. M. and Wrolstad R. E. (2005). Handbook of analytical food chemistry (pp. 19–31). Unit F1.2: Characterization and measurement of anthocyanins by UV–visible spectroscopy. In R. E. Wrolstad (Ed.), New York: John Wiley & Sons.

http://ijpjournal.com/bft-article/uv-visible-spectrophotometric-estimation-of-curcumin-in-nanoformulation/?view=fulltext

http://ijpjournal.com/bft-article/uv-visible-spectrophotometric-estimation-of-curcumin-in-nanoformulation/?view=fulltext

http://nopr.niscair.res.in/bitstream/123456789/11499/1/IJTK%2010%282%29%20 247-250.pdf

Iland, P., Ewart, A., Sitters, J., Markides, A. and Bruer, N. (2000) Techniques for chemical analysis and quality monitoring during winemaking. Patrick Iland Wine Promotions, Campbell town, SA.

Iland, P. G., Cynkar, W., Francis, I.L., Williams, P.J. and Coombe, B.G. (1996) Australian Journal of Grape and Wine Research; 2 171-178. Optimisation of methods for the determination of total and red free glycosyl-glucose in black grape berries of *Vitis vinifera.*

Khalil I. A. and F.R. Varananis. 1996. Sarhad J. Agric. 105 (67): 15-21.Carotiniod extraction and analysis by reversed phase HPLC system.

Michelle A Tait and David S. Hik (2003). Photosynthesis Research10- 2003, Volume 78, Issue 1, pp 87-91. Is dimethyl-sulfoxide a reliable solvent for extracting chlorophyll under field conditions?

Ranganna S (1976) In: Manual of Analysis of fruit and vegetable products. McGraw Hill New Delhi p. 77.

Sogi D.S., Sharama S Oberoi D. P. S. Wani I. A. (2010). J Food Sci Technol. 47, 3 pp 300-304. Effect of extraction parameters on curcumin yield from turmeric.

Wellburn AR (1994) J Plant Physiol144: 307-313. The spectral determination of chlorophyllsaandb, as well as total carotenoids, using various solvents with spectrophotometers of different resolution.

Witham F H., Blaydes D F and Devlin R M (1971) Expriments in plant physiology Van Nostrand New York p. 245.

Wrolstad R. E., Durst R. W. and Lee J. (2005). Trends in Food Science and Technology. 16. 423–428. Tracking color and pigment changes in anthocyanins products.

Yoder BJ and Daley LS (1989) Spectroscopy8 44-50. Development of a visible spectroscopic method for determining chlorophylla and in vivo in leaf samples.

9 Chapter

Analysis of Plant Growth Regulators

9.1 Principle

Plant growth regulator and polar compound residues are extracted from the grape with the addition of acidified methanol. The mixture is centrifuged, filtered and directly analyzed by LC MS/MS. Quantification is performed with the help of isotopically labeled analogues of the target analytes, which are used as standard (ISTD). These ISTDS are added directly to the sample at the beginning of procedure to compensate for any factors having an influence on the recovery losses during the sample preparation as well as matrix effects during measurement. Quantification and analysis is done by external standard calibration method.

9.2 Materials and Methods

- Blender or vertical cutter-mixer
- Vortex mixer
- Centrifuge
- Balance, accurate to 0.01 mg
- Rough balance, 0.001 to 420 g capacity.
- Centrifuge tubes- polypropylene, 50 ml and 15 ml for sample extraction and clean up-tarson.
- Suitable auto sampler vials
- Volumetric flasks- 10, 25, 50 ml.
- Micro centrifuge tubes- 2.0 ml from eppendrof
- Standard storage screw cap bottles with septa
- Micro- pipettes (variable volumes) 10, 100, 1000 and 5000μl capacity
- Solvent filtration assembly
- Nylon 66 membrane filter paper 10 mm 25 mm diameter with 0.2 μ pore size
- Sartorius water purification/ deionization system.

9.3 Instrumentation

a) Triple quadrupole linear ion trap mass spectrometer API 5500 Q-Trape – (Applied Biosystemsciex) equipped with electro spray ionization (ESI) and analyst 1.5.1. Data system. HPLC system

b) Triple quadrupole linear ion trape mass spectrometer API 4000 Q Trap- (Applied Biosystemsciex) equipped with electro spray (ESI) interface and Analyst 1.5.1 data system. HPLC system (Agilent 1200 series HPLC)

c) Triple quadrupol mass spectrometer API 2000 applied biosystemSciex) equipped with electro spray (ESI) LC interface and analyst 1.5 data system. HPLC system (Perkin Elmer 200 series)

9.4 Reagent and Chemicals

a) Methanol- HPLC gradient grade.

b) Water - HPLC gradient grade.

c) Formic acid (concentration > 95%)

d) Mobile phase B: 0.1 % formic acid in methanol and water: add 1 ml formic acid to 1000 ml in graduated cylinder. Dilute to 1000 ml with 900 ml methanol and 99 ml water. Filter through 0.2 μ N66 filter using solvent filtration assembly. Sonicate for 5 min. for degassing the phase.

e) Acidified methanol: add 1 ml of formic acid to 1000 ml graduated cylinder. Dilute to 1000 ml with methanol.

9.5 Standards

9.5.1 Reference Standards

Reference standards of Chlormequat chloride (CCC), D4 chlormiquat CI, Ethephon, D4 ethephon, Gibberellin A3, 2,4 D, 6- BA, Indole-3-Acetic acid (IAA) (Auxin), IBA, Zeatin (Cytokinin), kinetin, CPPU, Paclobutrazole, Isoprothiolane, Uracle, Abscic acid (ABA), were obtained from ehrenstorfereGmhh and homobrassinolide technical standard from local manufacture.

9.5.2 Standard Solutions

Take stock solution of individual PGR reference standards by accurately weighing 10 mg/ or (ml) of each analyte in volumetric flasks and dissolve in 10 ml of methanol. Cap tightly and mix and stored in dark amber coloured vials 4^0C.

9.5.2.1 Internal Standard Stock Preparation

A working standard mixture of 1 mg/L prepared by appropriate serial dilution of the stock solutions, from which the calibration standards (0.001–0.05 lg/mL) were prepared by serial dilutions with methanol: water (1:1, v/v). The solution of Ethephon D4 (IS) was diluted to 10 lg/mL with methanol.

9.5.2.2 External Calibration Solvent Standards for LC-MS/MS

An appropriate calibration range of pesticides working standard mixture is prepared. 0.0025 – 0.1 μg/ ml for LC-MS/MS API 4000 Q TRAP and 0.001 – 0.025 μg/ ml API 5500 Q -TRAP. Inject one intermediate standard prior to each day's run of sample to determine instrument suitability. Prepare fresh each day of use.

9.5.2.3 External Calibration Matrix Standards

An appropriate range of plant growth regulators mixed matrix calibration standard may be prepared by post extraction spike of the methanol extract a previously tested residue free sample.

9.5.2.4 Preparation of Controls (to be included as part of each sample batch)

Note: controls should batch to the samples to be confirmed.

a) Control samples are from plants to be free of pesticides. If these are not available, samples from an unknown sourced may be used provided it is first tested and show to be free of contamination.

b) Positive controls are control sample that have been fortified with known concentration of pesticides before contaminant.

9.6 Sample Preparation

9.6.1 Sample Handling and Preparation

a) Freshly collected samples (two no.) must be kept in cold before and during transporting it the laboratory. Once received at laboratory, samples must be frozen (-18°C) prior to mincing/ grinding if they cannot be prepared on the day of receipt.

b) If sample is frozen, allow defrosting, but keeping as cold as possible. Grind the berries in blender or vertical cutter- mixer. Stored frozen (-18°C) prior to analysis.

c) Crush the laboratory sample into small pieces (about 1 cm2) in a blender.

d) Take approximately 200 g of the crushed sample and homogenize it for one minute.

9.6.2 Digestion

a) Weigh 10.00 + 0.10 g ground the plant samples in a 50 ml polypropylene c entrifuge tube and kept for 10 min. (Note: prepare necessary controls at this time.)

b) Add 20 ml acidified methanol to the sample and 50 µl of the ISTD-WS containing isotopically labeled analog of one or more of the analyst of interest.

c) Close the tubes and shake/ vortex vigorously for 1 min.

d) Centrifuge at approximately 2000- 2500 rpm for 1 min.

e) Draw 2 ml supernatant in the eppendrof tube.

f) Centrifuge the extract at 1000 rpm for 5 min. and filter through 0.2 µm N66 membrane filter into the LC-MS/MS vial.

g) Inject 10 µl/ 20 µl from the extract into the LC-MS/MS.

h) For every batch of sample fallow the same extraction protocol to process one reagent blank sample and spiked as quality check.

9.6.3 Sequence of Injection

a) Inject the solvent blank before and after standards and after every six samples to check that the system is free is free from carry over.
b) Inject mixture of solvent standards / matrix standard at least 5 levels as per the instrument used.
c) Inject reagent blank and positive control sample as a quality check.
d) Inject samples
e) If there are more than ten samples in a batch, after every 10 sample inject one solvent blank and one calibration standard to check carry over and overall performance of the analytical instrument.

9.7 Analytical Procedure for LC-Ms/Ms

Note- flows and elution gradient may be optimized, if necessary, for the best separation and response and documented at the same time.

Follow the work instructions for the respective LC-MS/MS system for the chromatographic conditions and mass parameters for the analysis of pesticide residues as per the Annexure I.

9.8 Confirmation

9.8.1 Data Processing

Check the retention time and response of calibration standards is proper. Prepare a quantitation method using a mid concentration level. Using the quantitation method prepare the quatitation result file with the calibration standard. Check linearity anregression equation within range (> no.99) for the analyte. Save the quantitation result file indicating analysis type and date. Load the samples on to the quantitation result file. Add appropriate dilution factor. Check the samples with respect to analyte for proper base selection. Save the data. Prepare the analyte report for the confirmed result.

9.8.2 Confirmation of An Analyte in Any Sample Requires the Following:

Retention time of the analyte peak in one or more of the chromatograms must match that found in the chromatogram of the calibration of the standard within the same analysis set, within 0.2 minutes. The intensity of the compound in the extracted ion chromatogram must be of signal/ noise (s/n) ratio of at least 10.

9.8.3 Criteria for Repeating an Analysis

a) The instrument is suspected to be malfunctioning, as demonstrated by; clearly aberrant standard spectra; failure of a calibration check performed shortly after analysis of the sample set; instrumental parameters.
b) There is suspected carryover from a previously high concentration sample or standard. In this case, the sample should be reanalyzed after the cause of the carryover has been identified and measures taken to prevent its recurrence.

9.9 Analysis of Hydrogen Cyanamide by Calorimetric Method

9.9.1 Principle

Cyanamid precipitated by the addition of excess ammonical silver nitrate. The silver Cyanamid is dissolved in acid and titrated with this Cyanamide solution. Since one mole of thiourea reacts to form one mole of silver. Cyanamid and one of silver sulfide, the resulting interference may be corrected by dissolving silver sulphite formed and titrating with thiocyanate.

9.9.2 Regents

1. **Ammonical silver Nitrate (approximately 0.5 N):** Dissolve 85 gm of silver nitrate in 500 ml of water. Slowly add conc. Ammonium hydroxide until the brown precipitate which forms and is re- dissolved. Add 1 ml of excess and dilute to one litter of water.
2. **Potassium thiocyanate (0.1 N standardized):** Dissolved 9.72 g of KSCN (AR) in one liter of distilled water in 1 litter of volumetric flask.
3. **Ferric alam indicator (2%):** Dissolve 2 g ferric alum in 100 ml of distilled water.
4. **Nitric acid (1 N): Add** 10 ml of concn. $HNO_3$to 150 ml of distilled water.

9.9.3 Procedure

1. Take 5 to 7 gm of sample in 500 ml of volumetric flask.
2. Dilute with distilled water, mix and adjust to mark.
3. Pipette out 10 ml of aliquot in to 500 ml of beaker containing 10 ml of water.
4. Add 50 ml of ammonical silver nitrate and mix well.
5. Allow to stand for 15 min. with occasional stirring and decant the supernent liquid through a sintered galss crucible of medium porosity.
6. Discard the filtrate. Wash the ppt in a beaker several times with water containing a few drops of NH_4 OH by decanting the liquid until the washing is free of silver. (test by acidifying with hydrochloric acid , a ppt indicate that silver is present)
7. Dissolve the ppt in 30 ml of 1 N Nitric acid pure the solution through the same sintered glass crucible used above and wash well with distilled water.
8. Dilute the filter to approximately 125 ml of distilled water.
9. Add 1 ml of ferric alum indicator and titrate with 0.1 N KSCN (potassium thiocyanate)to a salmon pink end point.
10. If black ppt of silver sulfide remains after solution of silver cynamide, dissolved the ppt in 5 ml pf hot connc. Nitric acid, dilute to 125 ml and titrate with 0.1 N KSCN solutions.

9.9.4 Calculation

$$Hydrogen\ Cyanamide\ (H_2CN_2)\% = \frac{(A\text{-}B) \times N \times 500 \times 0.021 \times 100}{Wt.\ of\ sample \times 50}$$

Where,

Ml. of thiocyanate used to titrate silver cyanamide solution

Ml. Ofthiocyanate used to titrate silver cyanamide solution.

N- Normality of KSCN.

9.10 Analysis of Hydrogen Cyanamide by HPLC Method

9.10.1 Theory

The continuous improvement in HPLC columns and instrumentation presents an opportunity to improve HPLC methods. A proven ortho-phthalaldehyde/9-fluorenylmethyl chloroformate (OPA/FMOC) derivatized amino acid analysis method developed on the HP 1090 Series HPLC Systems, and later updated for the Series HPLC Systems, has now evolved further taking advantage of the Agilent 1200 Series SL and Agilent ZORBAX Eclipse Plus C18 stationary phase columns.

A chromatogram is presented for each column or method.

They are grouped into three categories:

- Traditional 5-μm particle size columns that offer high resolution
- Practical 3.5-μm particle size columns that combine high resolution and speed of the 5 and 1.8-μm particle size columns (Rapid Resolution)
- Highest throughput 1.8-μm particle size columns that offer (Rapid Resolution High Throughput) excellent resolution.

The 2.1 × 150 mm, 3.5 μm Rapid Resolution method is a preferred replacement method for the original AminoQuant method that used column p/n 79916AA-572.

All of the methods use the same chemicals and OPA/FMOC derivatization based on previous Agilent methods. The differences among these methods are the mobile phase gradientprograms and the flow path. These two parameters areoptimized according to column dimensions and LC instrumentmodels.

9.10.2 Sample Preparation

Amino acid standard mixtures and internal standards were obtained from Phenomenex and combined into a single standard mix for analysis. The standard mixtures contained the following 36 amino acids, each with a starting concentration of 200 nmol/mL.

Glutamine (GLN), Arginine (ARG), Serine (SER), Citrulline (CIT), 1-Methylhistidine (1-MHIS), 3-methylhistidine (3-MHIS), Glycine (GLY), 4-hydroxyproline (HYP), Glycine-prolinedipeptide (GPR), sarcosine (SAR), Alanine (ALA), gamma-aminobutyric acid (GABA), betaaminoisobutyric acid (BAIB), alpha-aminobutyric acid (ABA), Proline (PRO), Methionine (MET), Ornithine (ORN), Valine (VAL), Aspartic acid (ASP), Histidine (HIS), Glutamic acid (GLU), Lysine (LYS), Tryptophan (TRP), Leucine (LEU), Isoleucine (ILE), Phenylalanine (PHE), Aminopimelic acid (APA), Cystathionine (CTH), Cystine (C-C), Tyrosine (TYR), alpha-aminoadipic acid (AAA), Threonine (THR), Asparagine (ASN), delta-hydroxylysine (HLY), prolinehydroxyproline dipeptide (PHP), thiaproline (TPR).

The internal standard mix contained the following 3 amino acids each at a starting concentration of 200nmol/mL. Homoarginine (HARG), d3-methionine (d3-MET), homophenylalanine (HPHE) Sample preparation consisted of a fast SPE step (ion exchange in packed pipet tips) followed by derivatization with propyl chloroformate and subsequent clean-up by liquid-liquid extraction. The derivatization protocol resulted in amino acid species with greater chromatographic retention, resulting in improved separation of polar amino acids. The derivatized amino acids may also show improved ionization in +ESI due to blocking of acidic functional groups.

9.10.3 Chemical Preparation

9.10.3.1 Mobile Phase A

- 10 mM Na_2HPO_4: 10 mM $Na_2B_4O_7$, pH 8.2: 5 mM NaN_3 For 1 liter: (1.4 g anhydrous Na_2HPO_4 + 3.8 g $Na_2B_4O7{\cdot}10H_2O$ in 1 L water + 32 mg NaN^3). Adjust to about pH 8.4 with 1.2 mL concentrated HCl, and then add small drops until pH 8.2. Allow stirring time for complete dissolution of borate crystals before adjusting pH. Filter through 0.45-μm regenerated cellulose membranes.

9.10.3.2 Mobile Phase B Acetonitrile Methanol Water (454510, v v v).

- All mobile-phase solvents are HPLC grade. Since mobile phase A is consumed at a faster rate than B, it is convenient to make 2 L of A for every 1 L of B.

9.10.3.3 Injection Diluent

- 100 mL of mobile phase A + 0.4 mL concentrated H3PO4 in a 100-mL bottle. Store at 4°C.

9.10.3.4 0.5 N HCl

- Add 4.2 mL of concentrated HCl (36 %) to a 500-mL volumetric flask that is partially filled with water. Mix, and fill to mark with water. Solution is for making Extended Amino Acid and Internal Standard stock solutions. Store at 4°C.

9.10.3.5 Derivatization Reagents

- Borate buffers, OPA and FMOC, are ready-made solutions supplied by Agilent. They simply need to be transferred from their container into an autosampler vial. Some precautions include:
- OPA is shipped in ampoules under inert gas to prevent oxidation. Once opened, the OPA is potent for about 7-10 days. It is recommended to transfer 100-μL aliquots of OPA to micro vial inserts, label with name and date, cap, and refrigerate. Replace the OPA autosampler micro vial daily. Each ampoule lasts 10 days.
- FMOC is stable in dry air but deteriorates in moisture. It is also recommended to transfer 100-μL aliquots of FMOC to micro vial inserts, label with name and date, cap tightly, and refrigerate. Like the OPA, an

open FMOC ampoule transferred to 10 micro vial inserts should last 10 days (one vial/day).

- Borate buffer can be transferred to a 1.5-mL autosampler vial without a vial insert. It can be replaced every 10 days.

9.10.4 Preparing the Amino Acid Standards

9.10.4.1 Amino Acid Standards (10 pmol/µL to 1 nmol/µL):

- Solutions of 17 amino acids in five concentrations are available from Agilent for calibration curves. Divide each 1-mL ampoule of standards p/n 5061-3330 through 5061-3334 into 100-µL portions in conical vial inserts. Cap and refrigerate aliquots at 4 °C.

9.10.4.2 Extended Amino Acid (EAA) Stock Solution

- Weigh 59.45 mg asparagine, 59.00 mg hydroxyproline, 65.77 mg glutamine, and 91.95 mg tryptophan into a 25-mL volumetric flask. Fill halfway with 0.1 N HCL and shake or sonicate until dissolved. Fill to mark with water for a total concentration of 18 nmol/µL of each amino acid. For high-sensitivity EAA stock solution, take 5 mL of this standard-sensitivity solution and dilute with 45 mL water (1.8 nmol/ µL). Solutions containing extended standards are unstable at room temperature. Keep them frozen and discard at first signs of reduced intensity.

9.10.4.3 Internal standards (ISTD) Stock Solution

- For primary amino acids, weigh 58.58 mg norvaline into a 50-mL volumetric flask. For secondary amino acids, weigh 44.54 mg sarcosine into same 50-mL flask. Fill halfway with 0.1 N HCl and shake or sonicate until dissolved, then fill to mark with water for a final concentration of 10 nmol each amino acid/µL (standard sensitivity). For high-sensitivity ISTD stock solution, take 5 mL of standard sensitivity solution and dilute with 45 mL of water. Store at 4°C. Calibration curves may be made using two to five standards depending on experimental need. Typically 100pmol/µL, 250 pmol/µL, and 1 nmol/µL are used in a three-point calibration curve for "standard sensitivity" analysis. The following tables should be followed if internal standard or other amino acids (for example, the extended amino acids) are added. Table 2 describes "standard sensitivity" concentrations typically used in UV analysis.

9.10.4.4 Distinct Method Parameters

9.10.4.5 The LC pump

- Compressibility (×10-6 bar) A: 35, B: 80
- Minimal Stroke A, B: 20 µL

9.10.4.6 The automated Liquid Sampler (ALS): Online Derivatization

- Depending on the autosampler model, the automated online derivatization program differs slightly:

9.10.4.7 G1376C Well Plate Automatic Liquid Sampler (WPALS), with Injection Programme

1. Draw 2.5 μL from borate vial (Agilent p/n 5061-3339).
2. Draw 1.0 μL from sample vial.
3. Mix 3.5 μL in washport 5 times.
4. Wait 0.2 min.
5. Draw 0.5 μL from OPA vial (Agilent p/n 5061-3335).
6. Mix 4.0 μL in washport 10 times default speed.
7. Draw 0.4 μL from FMOC vial (Agilent p/n 5061-3337).
8. Mix 4.4 μL in washport 10 times default speed.
9. Draw 32 μL from injection diluent vial.
10. Mix 20 μL in washport 8 times.
11. Inject.
12. Wait 0.1 min.
13. Valve bypass.

9.10.4.8 G1329A Automatic Liquid Sampler (ALS), with Injection Program

1. Draw 2.5 μL from borate vial (Agilent p/n 5061-3339).
2. Draw 1.0 μL from sample vial.
3. Mix 3.5 μL in air default speed 5 times.
4. Wait 0.2 min.
5. Draw 0.5 μL from OPA vial (Agilent p/n 5061-3335).
6. Mix 4.0 μL in air, 10 times default speed.
7. Draw 0.4 μL from FMOC vial (Agilent p/n 5061-3337).
8. Mix 4.4 μL, 10 times default speed.
9. Draw 32 μL from injection diluent vial
10. Mix 20 μL in air default speed 8 times.
11. Inject.
12. Wait 0.1 min.
13. Valve bypass

The location of the derivatization reagents and samples is upto the analyst, and ALS tray configuration. Using the G1367Cwith a 2 × 56 well plate tray (G2258-44502), the locations were:

- Vial 1: Borate buffer
- Vial 2: OPA
- Vial 3: FMOC
- Vial 4: Injection Diluent
- P1-A-1: Sample

9.10.4.9 The Thermostatted Column Compartment (TCC)

Left and right temperatures are set at 40 °C. Enable analysis when temperature is within ± 0.8 °C. See Table 5 for which heat sink to use.

9.10.4.10 The Diode Array Detector (DAD)

9.10.4.10.1 Signal A

338 nm, 10 nm bandwidth, and reference wavelength 390 nm, 20 nm bandwidth.

9.10.4.10.2 Signal B

262 nm, 16 nm bandwidth, and reference wavelength 324 nm, 8 nm bandwidth.

9.10.4.10.3 Signal C

338 nm, 10 nm bandwidth, and reference wavelength390 nm, 20 nm bandwidth. Programmed to switch to262 nm, 16 nm bandwidth, reference wavelength 324 nm,8 nm bandwidth, after lysine elutes and before hydroxyproline elute. Signal C is determined by examining signal A and B time frames between peaks 20 and 21, then choosing a suitable point in time to switch wavelengths. Once switch time is established and programmed into the method, signal A and Bare optional.

9.10.4.10.4 Peak Width Settings Were

- >0.01 min for the Rapid Resolution High Through put(RRHT) 1.8-µm
- column methods
- >0.03 min for the Rapid Resolution (RR) 3.5-µm column methods
- >0.03 min for the Traditional High Resolution 5-µm column methods

9.10.4.10.5 Columns and Guard Cartridges

- See Table 5 for Agilent ZORBAX Eclipse Plus C18 columns and recommended guard cartridges. A Cartridge Hardware Kit (p/n 820888-901) is needed to house the guard cartridge. Cartridges should be flushed with a few milliliters of mobile phase B before connecting to column for equilibration.

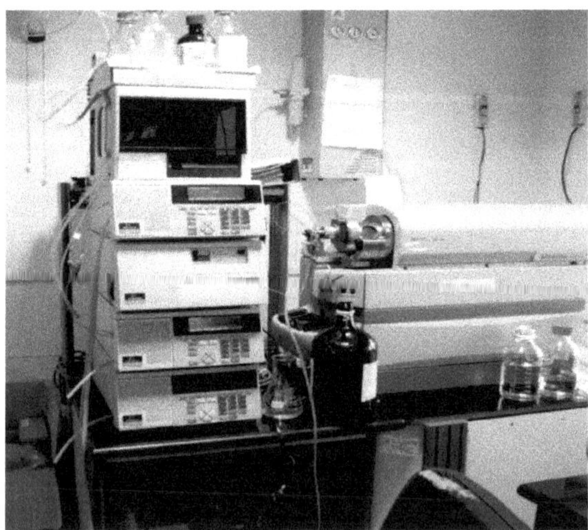

9.10.5 Results and Discussion

The scalability of Agilent ZORBAX Eclipse Plus C18 stationary phase columns between three particle sizes (5, 3.5, 1.8 µm) allows for many fine tuned methods. As summarized in Table 1, the analyst can choose a column or method based on analysis time, resolution factors, solvent usage, or instrument model. Each of the three categories has its own method parameters. Of particular importance is the mobile phase flow path configuration for the smaller volume columns. Truly scaling a gradient method between different column diameters and lengths is somewhat complicated, especially because different flow rates produce different gradient delay times, which is the time it takes the gradient to travel from its formation point to the column head.

Besides flow rate, the delay time is determined by the flow path volume between the gradient formation point and the head of the column. The delay volume can vary from instrument to instrument. The specific flow paths used to produce all the chromatograms in this application note are listed in Table 5. Instead of trying to exactly scale the methods, a simpler approach was used with the flow path volumes listed in

Table 5 and flow rates that produced a good separation. For example, the 4.6 × 50 mm, 1.8 um separation has a flow rate of 2 mL/min instead of 1. 5 mL /min, which is the equivalent flow rate, used for the longer 100, 150 and 250 mm, 4.6 mm id columns.

Workflow: HPLC/MS/MS analysis workflow

1. Prepare the HPLC system

a. Prepare Mobile Phase A and B

b. Set up the HPLC System

c. Connect to the mass spectrometer

2. Prepare the MS system

a. Perform a System Suitability Test

b. Review the test results

c. If necessary, update the acquisition and quantitation methods with the retention times

3. Perform the sample assay

a. Create a project folder

b. Load the autosampler

c. Perform the sample assay

References

Angelika Gratzfeld-Huesgen. (1999). Sensitive and Reliable Amino Acid Analysis in Protein Hydrolysatesusing the Agilent 1100 Series HPLC Agilent Pub. # 5968-5658EN

Bharat Ugare, Kaushik Banerjee, S. D. Ramteke, Saswati Pradhan, Dashrath P. Oulkar, Sagar *C. Utture and* Pandurang G. Adsule. (2013). Dissipation kinetics of forchlorfenuron, 6-benzyl aminopurine, gibberellic acid and ethephon residues in table grapes (*Vitis vinifera*). *Food Chemistry* 141, 4208-4214.

Cliff Woodward, John W Henderson Jr. and Todd Wielgos. (2007). Agilent Pub.# 5989-6297EN. High-Speed Amino Acid Analysis (AAA) on Sub-Two Micron Reversed-phase (RP) Columns.

Herbert Godel, Petra Seitz, and Martin Verhoef. (1992). LC-GC International, 5(2), 44-49.

Jason Greene, John W. Henderson Jr., John P. Wikswo, (2009). Agilent Pub. # 5990-3283EN. Rapid and Precise Determination of Cellular Amino Acid Flux Rates Using HPLC with Automated Derivatization with Absorbance Detection.

John W. Henderson Jr, Robert D. Ricker, Brian A. Bidlingmeyer, and Cliff Woodward, (2000). Agilent Pub. # 5980-1193E. Rapid, Accurate, Sensitive and Reproducible HPLC Analysis of Amino Acids.

Rainer Schuster (1988). *J. Chromatogr.*, 431, 271-284

Rainer Schuster and Alex Apfel (1986). Hewlett-Packard App. Note, Pub. # 5954-6257.

10

Chapter

Analysis of Macro and Micro-Nutrients From Plant Samples

Introduction

Plant nutrient analysis is based on the principle that the concentration of a nutrient within the plant is an integral value of all the factors that have interacted to affect it. Plant analysis involves the determination of nutrient concentration in diagnostic plant parts at recommended growth stage. The concentration of some nutrient elements may be too low for optimum growth, while of others may be so high as to be detrimental to the plant's growth, overviews such as that of Munson and Nelson (1990) illustrate how nutrient concentrations in plant vary with the element in question, type of plant, specific plant part, growth stage, level of available nutrients, expected yield level, and environmental factors. The knowledge of nutrient concentration in growing plants can serve as a tool for correcting any deficiencies if carried out early enough to safeguard yield. It can also be used to evaluate the efficacy of a recent application. Analysis of harvest plant parts or mature tissues is like to post mortem. The information it provides can help to plan nutrient application only in subsequent year on that field.

- **Plant Sampling:**

  Years of research in soil fertility- plant nutrition have produced reliable sampling criteria and procedures for most of the crops. Leaves are most commonly chosen, but petioles are selected in certain cases, seeds are rarely used for analysis, except for assessing of Boron toxicity and Zn and P Deficiency in certain grain crops. In some cases for example cereals the entire above ground young plants are sampled.

  When leaves are sampled, recently matured ones are taken; both new and old growth is generally avoided. However, young emerging leaves are sampled for diagnosed leaves are excluded, and plant should not be sampled when the crop is under moister or temperature stress.

  Plant samples should be transported to the laboratory immediately in properly labeled paper bags that allow for transpiration; this reduces the possibility of rooting.

- **Laboratory Processing:**

  After sample collection, the fresh tissue should be decontaminated from dust and other foreign material by adopting the following procedure.

1. The fresh tissue should be washed by de-ionized water or with 0.1- 0.3 % p- free detergent.
2. Sample should be washed with sequence in detergent solution, dilute HCL and deionized water. The liquid detergent will remove waxy coating on leaf surface and any soil particles.
3. N/ 10 HCL will remove metallic contaminants and deionized water will wash the previous two solutions.
4. Immediate drying in an oven to stop enzymatic activity, usually at 65^{0}C for 24 hours.
5. Mechanical grinding to produce a material suitable for analysis, usually to pass a 60- mesh sieve; stainless still mills are preferable, particularly when micro-nutrient analysis is involved.
6. Since most analytical methods require grinding of a dry sample, careful attention must be given to avoiding contamination with the element to be analyzed. Particularly care is required for micro-nutrients.
7. Final drying at 65^{0}C of ground tissue to obtain a constant weight upon which to base the analysis.

Methods of Analysis

The concentration of nutrients in plant tissues can be measured in a plant extract obtained from fresh plant material, (i.e., leaf & petiole analysis), as well as in whole dried plant material. The former test is qualitative and is appropriate only for quick measurements on a growing crop. Total plant analysis is quantitative in nature and is more reliable and useful. Of prime concern are forms of N, as well as P, B, and micronutrient cations. More detailed interpretative guidelines for plant analysis data are available in Reuter and Robinson (1986, 1997) and Jones *et al.* (1991).

10.1 Determination of Nitrogen

One common plant analysis is that of nitrogen (N) by Kjeldahl method. However, wet ashing with H_2SO_4 and H_2O_2 is also used for eliminating the use of selenium in the former method.

10.1.1 Kjeldahl Method for Determination of Nitrogen from Plant Sample

Principle

Nitro compounds formed by the reaction of salicylic acid with No 3 in acid medium are reduced to corresponding amino compounds by heating the mixture with sodium thiosulfate and zink dust. The main product of nitration is 5- nitrosalicylic acid and small amount of 3- nitrosalicylic acid. Drying of plant samples in oven eliminates interface of water in analysis as nitration of salicylic acid does not take place if more than a trace of water is present.

The methods involves the following important steps:

1. Sample preparation

2. Digestion of the sample to convert the N compounds in the sample to NH_4^+ from
3. Determination of NH_4^+ in the digest.

Apparatus

Block-digester

Distillation unit

Automatic titrator connected to a pH-meter

Vortex tube stirrer

Reagents

The chemicals used here are the same as for soil Kjeldahl-N.

- Catalyst Mixture (K_2SO_4-Se), 100: 1 w/ w ratio
- Sulfuric Acid (H_2SO_4), concentrated
- Ethylene Diaminetetraacetic Acid Disodium Salt (EDTA), M.W. = 372.2
- Sodium Hydroxide Solution (NaOH), 10 N
- Boric Acid Solution (H_3BO_3), saturated
- Sulfuric Acid Solution (H_2SO_4), 0.01 N
- G. Standard Stock Solution: 1.2 g NH_4 +-N per L.

A. Sample Preparation and Digestion

1. Fallow the seven steps of **Laboratory processing**
2. Weigh the 0.5 g of sample. However it varies from 0.2 to 2 g depending on the level of nitrogen.
3. Mix and spread finely ground (Cyclone mill) plant sample in a thin layer on a sheet of paper until it looks uniform.
4. Select representative sub-samples of about 1 g by taking at least 10 small portions from all parts of the sample with a spatula, and put them into a plastic vial.
5. Dry the sub-sample at 60°C in an oven (overnight), and then cool in a desiccator.
6. Weigh 0.25 g (grain) or 0.50 g (straw) of dry plant material, and transfer quantitatively into a 100-mL digestion tube.
7. Add a few pumice boiling granules, and add about 3 g catalyst mixtures using a calibrated spoon.
8. Add 10 mL concentrated sulfuric acid using a dispenser, and stir with Vortex tube stirrer until mixed well.
9. Place tubes in a block-digester set at 100°C for 20 minutes, and remove the tubes to wash down any material adhering to the neck of the tube with the same concentrated sulfuric acid. Thoroughly agitate the tube contents, and then place the tubes back on the block-digester set at 380°C for 2 hours after clearing.

10. After digestion is complete, remove tubes, cool, and bring to 100-mL volume with DI water.
11. Each batch of samples for digestion should contain at least one reagent blank (no plant), and one chemical standard (weigh 0.1 g EDTA standard digest), and one standard plant sample (internal reference).

B. Distillation

1. Set distillation and titration apparatus as for soil Kjeldahl-N, and steam out the apparatus for at least 10 minutes.
2. Prior to distillation, shake the digestion tube to thoroughly mix its contents. And pipette 10 mL aliquot into a 100-mL distillation flask.
3. Carefully add 10 mL 10 N sodium hydroxide solution, and immediately connect the flask to distillation unit and begin distillation.
4. Collect about 35 mL distillate in the collecting dish.
5. Remove distillation flask and connect an empty 100-mL distillation flask to the distillation unit. Drain water from the condenser jacket and steam out apparatus for 90 seconds before connecting the next sample.
6. Titrate the distillate to pH 5.0 with standardized 0.01 N H_2SO_4 using the Auto- Titrator; record titration volume of acid.
7. Each batch of distillations should include a distillation of 10 mL ammonium- N standard with 0.2 g MgO and 10 mL DI water with 0.2 g MgO. Recovery of ammonium-N standards should be at least 98%. Recovery of EDTA, corrected for reagent blank, should be at least 97%.

Calculations

- **Percentage recovery of Ammonium-N standard:**

$$\% \text{ Recovery} = \frac{(V - B) \times N \times 14.01 \times 100}{C \times D}$$

Where:

V = Volume of 0.01 N H_2SO_4 titrated for the sample (mL).

B = Distillate blank titration volume (mL)

N = Normality of H_2SO_4 solution.

C = Volume of NH_4-N standard solution (mL)

D = Concentration of NH_4-N standard solution (ìg/mL)

14.01= Atomic weight of N.

Percentage recovery of EDTA standard:

$$\% \text{ Recovery} = \frac{(V - B1) \times N \times R \times 186.1 \times 100}{Wt1 \times 1000}$$

- **Percentage Nitrogen in plant:**

$$\% N = \frac{(V - B1) \times N \times R \times 14.01 \times 100}{Wt2 \times 1000}$$

Where:

R = Ratio between total digest volume and distillation volume.

B1 = Digested blank titration volume (mL)

Wt 1 = Weight of EDTA (g)

Wt 2 = Weight of dry plant (g)

= Equivalent weight of EDTA.

10.1.2. Colorimetric Method for Determination of Nitrogen

Acolorimetric method was devised by Baethgen and alley (1989) for determination of NH_4-N in plant digest. This method can also be followed either manually or by using auto analyzer system.

Digestion

The digestion of sample is carried out as describe above, however, a different digestion mixture is used, and it is prepared by adding 30 parts of K_2SO_4+ 1 parts of H_2O as catalyst. Approximate 1 g digestion mixture is added to each tube.

Reagents

- Working buffer solution. (0.1 M Na_2HPO_4, 5 % Na- K tartarate , 5,4 % NaOH). Dissolve 26.8 g of $Na_2 HPO_4$. 7 H_2O in 600 ml of ammonium free water. Add 50 g Na- K tartarate and 108 g 50% (w/w) solution of NaOH. Dilute to 1 liter with distilled water.
- Na- Salicylate Na- Nitroprusside solution 1.5 % - 0.03 %). Dissolve 150 g Na- Salicylate and 0.30 g Na- Nitroprusside in 1 litre of distilled water. Store in a light resistant bottle.
- Na- hypochlorite solution. Dissolve 4.715 g dry (NH_4) SO_4 in a liter of distilled water. Prepare daily.
- Nitrogen standard solution. Dissolved 4. 715 g dry (NH_4) SO_4 in a liter of distilled water to prepare 1000 mg N/1. From this standard, solution containing 5.0, 10.0, 15.0, 30.0, 40.0, and 50.0 mg N/1 are prepared.

10.1.2. a. Determination of Nitrogen

1. Take 25 ml of volumetric flask in that add 1 ml of aliquot from plant digest.
2. Add 5.5 ml of solution 1 with constant stirring. Then add 4 ml of solution 2 and mix properly.
3. Add 1 ml of solution 3, mix it properly again. (solid residue should not be form)
4. The solution is allowed to stand for 45 minutes at 25°C (or 15 minutes at 37 °C) to ensure complete color development.
5. Take the absorbance at 650 nm in a spectrophotometer.

10.1.2 b. Determination of Total Nitrogen

This method is based on digestion of plant material in a sulfuric-salicylic acid mixture (Buresh *et al.*, 1982).

Reagents

- Sulfuric-Salicylic Acid Mixture (concentrated H_2SO_4 containing 2.5 % w/v salicylic acid)
- Dissolve 62.5g reagent-grade salicylic acid ($C_7H_6O_3$) in 2.5-L concentrated sulfuric acid.
- Catalyst Mixture (K_2SO_4-Se), 100:1 w/w ratio.
- Sodium Thiosulfate ($Na_2S_2O_3.5H_2O$), crystal
- Ethylene Di-amine tetra acetic Acid Disodium Salt (EDTA), M.W. = 372.2

Procedure

A Digestion

1. Mix and spread finely ground plant sample in a thin layer, on a sheet of paper or plastic until the sample looks uniform.
2. Take a representative sub-sample of about 1 g by systematically withdrawing at least 10 small portions from all parts of the sample with a spatula, and put them into a plastic vial.
3. Dry the sub-sample at 60°C in an oven (overnight), and then cool in a desic- desiccator.
4. Weigh 0.25 g (grain) or 0.50 g (straw) dry plant material, and then transfer quantitatively into a dry 250-mL digestion tube.
5. Add 20 mL sulfuric-salicylic acid mixture while rotating the tube to wash down any sample adhering to the neck of the tube, and allow standing 2 hours or longer with occasional swirling.
6. Add 2.5 g sodium thiosulfate through a long-stemmed funnel to the contents of the tube and swirl gently a few times, and allow standing overnight.
7. Add 4 g catalyst mixture, and 3-4 pumice boiling granules, and place tubes on the block-digester pre-heated to 400°C.
8. Place a small glass funnel in the mouth of the tubes to ensure efficient refluxing of the digestion mixture and prevent loss of H_2SO_4, and proceed with the digestion until the mixture clears.
9. Remove the tubes from the block-digester and allow them to cool for about 20 minutes. Then wash down any material adhering to the neck of the tube with a minimum quantity of DI water.
10. Thoroughly agitate the tube contents, place tubes back on the block-digester, and digest for 2 hours after clearing. No particulate material should remain in the tube after digestion.

11. After the digestion is finished, allow the digest to cool, and add water slowly shaking until the liquid level is about 2 cm below the graduation mark.
12. Allow tube to cool to room temperature, and add DI water to bring the volume to the 250 mL mark.
13. Each batch of samples for digestion should contain at least one reagent blank (no plant), and one chemical standard (weigh 0.1g EDTA standard digest), and one standard plant sample (internal reference)

B. Distillation

The reagents needed for distillation are the same as for soil Kjeldahl-N.

1. Set distillation and titration apparatus as for soil Kjeldahl-N, and steam out the apparatus for at least 10 minutes.
2. Prior to distillation, shake the digestion tube to thoroughly mix its contents, and pipette an aliquot in a 300-mL distillation flask.
3. Carefully add 7 mL or 15 mL 10 N sodium hydroxide solution for 25 mL or 50 mL aliquot, respectively, and immediately connect flask to distillation unit and begin distillation.
4. Collect about 35 mL distillate in the collecting dish.
5. Remove distillation flask and connect an empty 100-mL distillation flask to the distillation unit. Drain water from the condenser jacket and steam out apparatus for 90 seconds before connecting the next sample.
6. The distillate is then titrated to pH 5.0 with standardized 0.01 N H_2SO_4 using the Auto-Titrator; record titration volume of acid.
7. Each batch of distillations should contain at least two standards and two blanks (reagent blanks). Recovery of EDTA, corrected for reagent blank, should be at least 97%.

Calculations

$$\%\ \text{Recovery} = \frac{(V - B) \times N \times R \times 186.1 \times 100}{Wt1 \times 1000}$$

$$\%\ N = \frac{(V - B) \times N \times R \times 14.01 \times 100}{Wt2 \times 1000}$$

Where: V = Volume of 0.01 N H_2SO_4 titrated for the sample (mL).

B = Digested blank titration volume (mL)

N = Normality of H_2SO_4 solution.

14.01= Atomic weight of N.

R = Ratio between total digest volume and distillation volume.

Wt 1 = Weight of EDTA (g)

Wt 2 = Weight of dry plant (g)

186.1= Equivalent weight of the EDTA.

10.2 Determination of Phosphorus

Total P in plant material can be determined either by wet digestion procedure or by dry ashing procedure. Both methods are satisfactory. However, dry ashing is a simpler, easier, non-hazardous and economical option. Later, P content in the digests or dissolved ash aliquots is measured colorimetrically.

Apparatus

- Spectrophotometer or colorimeter, 410-nm wavelength
- Block-digester.
- Vortex tube stirrer.

Reagents

A. Ammonium Heptamolybdate-Ammonium Vanadate in Nitric Acid

- Dissolve 22.5 g ammonium hepta-molybdate [$(NH_4)6MO_7O_{24}$. $4H_2O$] in 400 ml DI water (a).
- Dissolve 1.25 g ammonium metavanadate (NH_4VO_3) in 300 ml hot DI water (b).
- Add (b) to (a) in a 1-L volumetric flask, and let the mixture cool to room temperature.
- Slowly add 250 mL concentrated nitric acid (HNO_3) to the mixture, cool the solution to room temperature, and bring to 1-L volume with DI water.

B. Standard Stock Solution

- Dry about 2.5 g potassium dihydrogen phosphates (KH_2PO_4) in an oven at 105°C for 1 hour cool in desiccator, and store in a tightly stoppered bottle.
- Dissolve 0.2197 g dried potassium dihydrogen phosphate in DI water, and bring to 1-L volume with DI water. This solution contains 50 ppm P.

C. (Stock Solution)

- Prepare a series of Standard Solutions from the Stock Solution as follows:
  - Dilute 1, 2, 3, 4, and 5 mL Stock Solution to100-mL final volume by adding DI water. These solutions contain 0.5, 1.0, 1.5, 2.0, and 2.5 ppm P, respectively.

Procedure

A. Wet-Digestion Procedure

1. Digest the plant material
2. Filter plant digest with Whatman No.1 filter paper, and collect filtrate in a small bottle. or as an alternative procedure

B. Dry-Ashing Procedure

1. Dry ash the plant material
2. Dissolve the ash in 2 N HCl.

C. Measurement

1. Pipette 10 ml of the digest filtrate or aliquot of the dissolved ash (depending on the procedure used) into a 100-mL volumetric flask, add

10 ml ammonium vanadomolybdate reagent, and dilute the solution to volume with DI water.

D. Prepare a Standard Curve as Follows

1. Pipette 1, 2, 3, 4, and 5 ml standard stock solution, and proceed as for the samples.
2. Also make a blank with 10 ml ammonium-vanadomolybdate reagent, and proceed as for the samples.
3. Read the absorbance of the blank, standards, and samples after 30 minutes at 410-nm wavelength.
4. Prepare a calibration curve for standards, plotting absorbance against the respective P concentrations.
5. Read P concentration in the unknown samples from the calibration curve.

Calculations

Percentage Total Phosphorus in plant

$$\% \text{ P} = \text{ppm P } (\text{from calibration curve}) \times \frac{R}{Wt} \times \frac{100}{1000}$$

Where:

R = Ratio between total volume of the digest/aliquot and the digest /aliquot volume used for measurement

Wt = Weight of dry plant (g)

Notes

The plant digest by the hydrogen peroxide and sulfuric acid can also be used for phosphorus measurement in plants.

10.3 Determination of Potassium

Principle

The most commonly used methods for determination of potassium is by flame photometery. It is based on the principle that atoms of some specific element take energy from flame and get excited to the higher orbit. Such a atom release energy of a wavelength which is specific for that element and is proportion to the concentration of atoms of that element.

Materials and Methods

Instrumentations

Theelement to be determined is introduced into the flame photometer in solution form. A fine aerosol is formed and the atom get excited by taking energy from flame created by mixture of liquid petroleum gas mixed with air. The emitted radiation may be several wavelengths. The isolation of radiation can also be achieved by a prism or monochromater. Radiation can measure either by photocell or photomultiplier tube. The concentration of K is measured by comparing the radiation emitted by a known standard with that of the sample.

Standard Stock Solution

To prepare a stock solution, 1.9069 g of analytical grade KCl is dissolved in DI water and make up the volume up to 1 liter. This solution contains 1000 ppm K. Prepare a 100 ppm K solution by diluting the 1000 ppm K solution 10 times (10 ml in 100ml of final volume). Final standard solutions of 0, 5 and 10 ppm are prepared from 100 ppm K.

Standard Curve

To prepare the standard curve, the instrument is set at highest concentration of 5 ppm using the standard filter. The manufactures specify the linear range of K (normally 5 ppm) and suitable factor is calculated for finding out K in plant sample.

Procedure

Estimation of K in Plant Samples

The plant samples for K estimation can be digested by diacid through wet ashing. The digest is diluted to suitable concentration range so that final concentration lies between 0 to 5 ppm. The samples are then read in flame photometer at 548nm wavelength or using filter for K.

Determination of water Soluble Potassium:

$$\text{K in \%} = R \times \frac{5}{100} \times \frac{100}{\text{sample}} \times \frac{100}{1000000}$$

Where

5 ppm K= 100 R and if further dilution is made, then appropriate modification is required in the calculation.

References

Estimation of Macro- and Micro-nutrients by Dry Ashing

- Plant analysis by dry ashing is simple, non-hazardous and less expensive, compared with HNO_3-$HClO_4$ wet digestion.
- Dry ashing is appropriate for analyzing P, K, Ca, Mg, and Na. Micro-nutrient cations (Fe, Zn, Cu, and Mn) can also be analyzed by dry ashing, but only in plant tissues containing low silica contents (like legumes).
- The HNO_3-$HClO_4$ wet digestion is required for full recovery of micronutrient cations in high-silica plant tissues (like wheat, barley, rice, and sugarcane, etc.).
- In dry ashing for B, use of glassware should be avoided

Materials and Methods

- Spectrophotometer or colorimeter, 410-nm wavelength
- Flame photometer
- Atomic absorption spectrophotometer
- Porcelain crucibles or Pyrex glass beakers (30 - 50 mL capacity)

Reagents

- Hydrochloric Acid (HCl), 2N
- Dilute165.6 ml concentrated hydrochloric acid (37%, sp.gr.1.19) in DI water, mix well, let it cool, and bring to 1-L volume with DI water.

Procedure

The procedure is that of Chapman and Pratt (1961) with slight modifications.

1. Weigh 0.5 - 1.0 g portions of ground plant material in a 30 - 50 mL porcelain crucibles or Pyrex glass beakers.
2. Place porcelain crucibles into a cool muffle furnace, and increase temperature gradually to 550°C.
3. Continue ashing for 5 hours after attaining 550°C.
4. Shut off the muffle furnace and open the door cautiously for rapid cooling.
5. When cool, take out the porcelain crucibles carefully.
6. Dissolve the cooled ash in 5-mL portions 2 N hydrochloric acid (HCl) and mix with a plastic rod.
7. After 15 - 20 minutes, make up the volume (usually to 50 mL) using DI water.
8. Mix thoroughly; allow to stand for about 30 minute. Then use the supernatant or filter through Whatman No. 42 filter paper, discarding the first portions of the filtrates.
9. Analyze the aliquots for P by Colorimetry (by Ammonium Vanadate-Ammonium Molybdate yellow color method), for K and Na by Flame Photometry, and for Ca, Mg, Zn, Cu, Fe, and Mn by Atomic Absorption Spectroscopy.

Notes

For Ca and Mg measurement, the final dilution should contain 1% w/v lanthanum (La) and the determinations should be against standards and blank containing similar La concentration to overcome anionic interference.

10.4 Determination of Boron

Boron in plant samples is measured by dry ashing (Chapman and Pratt, 1961) and subsequent measurement of B by colorimetry using Azomethine-H (Bingham, 1982).

10.4.1 Dry Ash Method

Materials and Methods

Apparatus

- Porcelain crucibles
- Spectrophotometer or colorimeter, 420-nm wavelength Polypropylene test tubes, 10 mL capacity

Reagents

a) Sulfuric Acid (H_2SO_4), 0.36 N

b) Buffer Solution

c) Azomethine-H

d) Standard Stock Solution

Procedure

A. Dry Ashing

1. Weigh 1 g dry, ground plant material in porcelain crucible.
2. Ignite in a muffle furnace by slowly raising the temperature to 550°C.
3. Continue ashing for 6 hours after attaining 550°C.
4. Wet the ash with five drops DI water, and then add 10 ml 0.36 N sulfuric acid solution into the porcelain crucibles.
5. Let stand at room temperature for 1 hour, stirring occasionally with a plastic rod to break up ash.
6. Filter through Whatman No.1 filter paper into a 50-mL polypropylene volumetric flask and bring to volume. Filtrate is ready for B determination.

B. Measurement

Same as in hot water extractable B in soils.

Calculations

For Boron in plant:

$$B\ (\text{ppm}) = \text{ppm B}\ (\text{from calibration curve}) \times \frac{A}{Wt}$$

Where:

A = Total volume of the extract (ml)

Wt = Weight of dry plant (g)

10.4.2 Determination of Boron by Hot-Water Method

The hot-water extraction procedure was introduced by Berger and Truog (1939), and was modified by later researchers. It is still the most popular method for measuring "available" soil B or the fraction of B related to plant growth in alkaline soils. Boron in soil extracts is measured colorimetrically using azomethine-H (Bingham, 1982). Where soil B levels are less than 0.5 ppm, deficiency is likely to occur for most crops. However, where levels are greater than about 5 ppm, toxicity may occur.

Materials and Methods

Apparatus

- Erlenmeyer flasks 50 mL (Pyrex), pre-treated with concentrated HCl for one week.

- Spectrophotometer or colorimeter, 420-nm wavelength
- Polypropylene test tubes, 10-mL capacity

Reagents

A. Buffer Solution

Dissolve 250 g ammonium acetate (NH4OAc), and 15 g ethylene diaminetetra acetic acid, disodium salt (EDTA disodium) in 400 mL DI water. Slowly add 125 mL glacial acetic acid (CH_3COOH), and mix well.

B. Activated Charcoal (Boron-free)

This is prepared by giving repeated washings (8 - 9 times) of DI water (boiling charcoal with water in 1:5 ratio), and subsequent filtering. Boron in the filtrate is checked by azomethine-H color development. Continue washing until it is B free.

C. Azomethine-H Solution ($C_{17}H_{12}NNa\ O_8S_2$)

Dissolve 0.45 g azomethine-H in 100 mL 1% L-ascorbic acid solution. Fresh reagent should be prepared weekly and stored in a refrigerator.

D. Standard Stock Solution

a) Stock Solution

Dissolve 0.114 g boric acid (H_3BO_3) in DI water, and bring to 1-L volume with DI water. This solution contains 20 ppm B.

Prepare a series of Standard Solutions from the Stock Solution as follows:

b) Dilute 2.5, 5.0, 7.5, 10.0, 12.5 and 15.0 mL Stock Solution to 100 Ml final volume by adding DI water. These solutions contain 0.5, 1.0, 1.5, 2.0, 2.5, and 3.0 ppm, respectively.

Procedure

A. Extraction

1. Weigh 10 g air-dry soil (2-mm) into a 50-mL Erlenmeyer flask (Pyrex), pre-treated with concentrated HCl for one week.
2. Add about 0.2 g activated charcoal (B-free).
3. Add 20 mL DI water.
4. Boil on a hot plate for 5 minutes with flask covered by a watch glass.
5. Filter the suspension immediately through Whatman No. 40 filter paper.

Filtrate is ready for B determination.

B. Measurement

1. Pipette 1 mL aliquot of the extract into a 10-mL polypropylene tube.
2. Add 2 mL buffer solution.
3. Add 2 mL azomethine-H solution, and mix well.
4. Prepare a standard curve as follows:

- Pipette 1 mL of each standard (0.5 - 3.0 ppm), and proceed as for the samples.
- Also make a blank with 1 mL DI water, and proceed as for the samples.
- Read the absorbance of blank, standards, and samples after 30 minutes at 420-nm wavelength.

5. Prepare a calibration curve for standards, plotting absorbance against the respective B concentrations.
6. Read B concentration in the unknown samples from the calibration curve.

$$\text{B (ppm)} = \text{ppm B (from calibration curve)} \times \frac{A}{Wt}$$

Calculations

For Extractable Boron in plants:

Where:

A = Total volume of the extract (mL).

Wt = Weight of air-dry soil (g)

Notes

Use of glassware should be minimal; and always use concentrated HCl-treated glassware (soaking for a week) where absolutely essential.

10.4.3 Dilute Hydrochloric Acid Method

Though the hot-water extraction method (HWE) is quite popular for predicting B availability in alkaline soils, the procedure is tedious and prone to error (because of difficulty in maintaining uniform boiling time). In an effort of having a convenient substitute, researchers (Kausar *et al.*, 1990; Rashid *et al.*, 1994; Rashid *et al.*, 1997) have found that the dilute HCl method of Ponnamperuma *et al.* (1981), originally designed for acid soils, is equally effective in diagnosing B deficiency in alkaline and calcareous soils. The HCl method is simple, economical, and more efficient.

Reagents

a) Buffer Solution

(Prepare as for hot-water extractable B).

b) Azomethine-H Solution ($C_{17}H_{12}NNa\,O_8S_2$)

(Prepare as for hot-water extractable B)

c) Activated Charcoal (Boron-free)

(Prepare as for hot-water extractable B).

d) Standard Stock Solution

(Prepare as for hot-water extractable B).

Hydrochloric Acid (HCl), 0.05 N

(Dilute 4.14 mL concentrated hydrochloric acid (37%, sp. gr. 1.19) in DI water, mix well, and bring to 1-L volume with DI water).

Procedure

A. Extraction

1. Weight 10 g air-dry soil (2-mm) into a polypropylene tube.
2. Add about 0.2 g activated charcoal (B-free)
3. Add 20 mL 0.05 N Hydrochloric acid solution.
4. Shake for 5 minutes, and then filter.

B. Measurement (by Azomethine-H Method)

$$B\ (ppm) = ppm\ B\ (\text{from calibration curve}) \times \frac{A}{Wt}$$

Calculations

Where:

A = Total volume of the extract (mL).

Wt = Weight of air-dry soil (g)

10.5 Determination of Chloride

Soluble chloride is obtained in the saturation extract (as prepared for soluble Ca, Mg and anions), and its concentration in the extract is determined by silver nitrate titration (Richards, 1954).

Material and Methods

Reagents

A. Potassium Chromate Solution (K_2CrO_4), 5% in Water

- Dissolve 5 g potassium chromate in 50 ml DI water.
- Add drop wise 1 N silver nitrate ($AgNO_3$) until a slight permanent red precipitate is formed.
- Filter, and bring to 100-mL volume with DI water.

B. Silver Nitrate Solution ($AgNO_3$), 0.01 N

- Dry about 3 g silver nitrate in an oven at 105°C for 2 hours, cool in a desiccators, and store in a tightly stoppered bottle
- Dissolve 1.696 g dried silver nitrate in DI water, and bring to 1-L volume with DI water.

C. Sodium Chloride Solution (NaCl), 0.01 N

- Dissolve 0.585 g dried sodium chloride in DI water, and bring to 1-L volume with DI water.

Procedure

1. Pipette 5 - 10 mL soil saturation extract into a wide-mouth porcelain crucible or a 150-mL Erlenmeyer flask.

2. Add 4 drops potassium chromate solution.
3. Titrate against silver nitrate solution until a permanent reddish-brown color appears.
4. Always run two blanks containing all reagents but no soil, and treat them in exactly the same way as for the samples. Subtract the blank titration reading from the readings for all samples.

Calculations

$$Cl\ (meq/L) = \frac{(V - B) \times N \times R \times 1000}{Wt}$$

For Chloride in soil:

Where:

V = Volume of 0.01 N $AgNO_3$ titrated for the sample (ml).

B = Blank titration volume (ml)

R = Ratio between total volume of the extract and extract volume used for titration.

N = Normality of $AgNO_3$ solution.

Wt = Weight of air-dry soil (g)

Standardization of $AgNO_3$

Titrate 10 mL 0.01N of sodium chloride solution against 0.01 N silver nitrate solution after adding 4 drops potassium chromate solution until a permanent reddish-brown color appears.

$$N\ AgNO_3 = \frac{10 \times N\ NaCl}{V\ AgNO_3}$$

Take the reading, and calculate $AgNO_3$ normality:

Where:

N $AgNO_3$ = Normality of $AgNO_3$ solution.

N $AgNO_3$ = Amount of $AgNO_3$ solution used (ml)

N NaCl = Normality of NaCl solution.

10.6 Determination of Sulfate

The commonly used method for sulfur (S) determination in alkaline soils is the extraction of SO_4-S with 0.15% $CaCl_2.2H_2O$ (Williams and Steinbergs, 1959) and measurement of SO_4-S concentration in the extracts by a turbidimetric procedure using barium chloride (Verma, 1977).

A critical range of 10 - 13 mg/kg $CaCl_2$-extractable SO_4-S has commonly been reported for cereal (e.g., wheat, maize) and oilseed (e.g., mustard) crops (Tandon, 1991).

10.6.1 Turbid Metric Method

Materials and Methods

Apparatus

- Mechanical shaker, reciprocal
- Spectrophotometer or colorimeter, 470-nm wavelength

Reagents

a) Calcium Chloride Dihydrate Solution ($CaCl_2.2H_2O$), 0.15%

Dissolve 1.5 g calcium chloride dihydrate in about 700 mL DI water, and bring to 1-L volumetric flask with DI water.

b) Hydrochloric Acid Solution (HCl), 6 M

Dilute 496.8 mL concentrated hydrochloric acid (37%, sp. gr. 1.19) in DI water, mix well, let it cool, and bring to 1-L volume with DI water.

c) Barium Chloride ($BaCl_2.2H_2O$), crystal

d) Sorbitol, 70% aqueous solution

e) Standard Stock Solution

Dissolve 0.5434 g potassium sulfate (K_2SO_4) in DI water, and bring to 1- L volume with DI water. This solution contains 100 ppm SO_4-S

f) Stock Solution

Prepare a series of Standard Solutions from the Stock Solution as follows:

Dilute the 5, 10, 20, 30, 40, and 50 ml Stock Solution to 100 ml final volume by adding 0.15% calcium chloride dihydrate solution. These standards contain 5, 10, 20, 30, 40 and 50 ppm SO4-S, respectively.

Procedure

A. Extraction

1. Weigh 5 g air-dry soil (2-mm) into a 150-mL Erlenmeyer flask.
2. Add 25 mL 0.15% calcium chloride dihydrate solution (don't use a rubber stopper, or wrap the rubber stopper in thin polyethylene. Errors result from gradual oxidation of sulfur compounds present in the stopper).
3. Shake for 30 minutes on a reciprocal shaker (180+ oscillations per minute).
4. Filter the suspension through Whatman No. 42 filter paper. This procedure yields almost colorless extracts.

B. Measurement of SO_4-S

1. Pipette 10-mL aliquot of the extract into a 50-mL test tube, or a smaller aliquot diluted to 10 mL with DI water.
2. Add 1 mL 6 M hydrochloric acid solution followed by 5 mL 70% sorbitol solution from a pipette with an enlarged jet. Finally, add about 1 g barium chloride crystals (using a measuring spoon).

3. Shake vigorously (on a test tube shaker for 30 seconds) to dissolve the barium chloride and obtain a homogeneous suspension.
4. Prepare a standard curve as follows:
   - Pipette 10 mL of each standard (0 - 50 ppm), and proceed as for the samples.
   - Also make a blank with 10 mL 0.15% calcium chloride dihydrate solution, and proceed as for the samples.
   - Read the absorbance (turbidity) of the blank, standards, and samples at 470-nm wavelength.
   - Prepare a calibration curve for standards, plotting absorbance against the respective SO_4-S concentrations.
   - Read SO_4-S concentration in the unknown samples from the calibration curve

Calculations

$$\text{SO4 - S}\left(\text{ppm}\right)=\text{ppm SO4 - S}\left(\text{from calibration curve}\right)\times\frac{A}{Wt}$$

For Turbidimetric of Sulfate in soil:

Where: A = Total volume of the extract (mL)

Wt = Weight of air-dry soil (g)

Note

1. Do not let the standards and unknowns (soil extracts) stand for longer than 2 - 3 minutes, otherwise re-shake the suspension before spectrophotometric reading.
2. Allow approximately the same time to standards and unknowns between shaking and turbidimetric reading.

10.6.2 Precipitation Method

Sulfate in water is determined normally by barium sulfate precipitation (Richards, 1954).

Apparatus

Mechanical shaker, reciprocating.

Muffle furnace.

Reagents

a) Methyl Orange Indicator [4-$NaOSO_2C_6H_4N$: NC_6H_4 /-4-N $(CH_3)2$], 0.1 %

   Dissolve 0.1 g methyl orange indicator in 100 mL DI water.

b) Hydrochloric Acid Solution (HCl), 1:1

   Mix equal portions of concentrated hydrochloric acid with DI water.

c) Barium Chloride Solution ($BaCl_2.2H_2O$), 1 N

Dissolve 122 g barium chloride in DI water, and bring to 1-L volume with DI water

Procedure

1. Put an aliquot of soil extract containing 0.05 to 0.5 meq SO_4-S into a 250-mL Pyrex beaker and dilute to 50 mL.
2. Add 1 mL 1:1 hydrochloric acid solution and 2 - 3 drops methyl orange; if the color does not turn pink, add some more 1:1 hydrochloric acid.
3. Put beakers on a hotplate, heat to boiling, then add 10 mL 1 N barium chloride solution in excess to precipitate SO_4 as barium sulfate.
4. Boil for 5 to 10 minutes, cover with a watchglass, and leave to cool.
5. Filter solution through ashless filter paper, collect the barium sulfate precipitate on the filter paper, and then wash it several times with warm DI water until no trace of chloride remains. The presence of chloride in the filtrate can be checked by $AgNO_3$ solution.
6. After washing, place filter paper with precipitate into a pre-weighed and dried porcelain crucible and put in an oven at 105°C for 1 hour to dry.
7. Transfer crucible to a muffle furnace heated to 550°C, and leave to dry ash for 2 - 3 hours.
8. Take crucible out of the muffle furnace, and place in a desiccator to cool, weigh crucible on an analytical balance, and take the reading, t.

$$SO_4\text{-}S\ (meq/L) = \times \frac{t\text{-}b}{V} 8583.7$$

Calculation

For precipitate of Sulfate in soil:

Where: t = Weight of crucible + $BaSO_4$ precipitate (g)

b = Weight of empty crucible (g)

V = Volume of extract used for measurement (mL)

10.7 Determination of Carbonate and Biocarbonate

Carbonate and bicarbonate are generally determined in soil saturation extract by titration with 0.01 N H_2SO_4 to pH 8.3 and 4.5, respectively (Richards, 1954).

Reagents

a) Methyl Orange Indicator [4-$NaOSO_2C_6H_4N{:}NC_6H_4$/-4-N (CH_3)2], (F.W. 327.34), 0.1%

Dissolve 0.1 g methyl orange indicator in 100 mL DI water.

b) B. Sulfuric Acid Solution (H_2SO_4), 0.01 N

- Dilute 28 mL concentrated sulfuric acid (98 %, sp.gr.1.84) in DI water, mix well, let it cool, and bring to 1-L volume with DI water.

This solution contains 1 N H_2SO_4 solution.

- Then dilute 100 times (10 mL to 1-L volume) to obtain 0.01 N H_2SO_4 solution.

c) Phenolphthalein Indicator, 1%

Dissolve 1 g phenolphthalein indicator in 100 mL ethanol.

Procedure

1. Pipette 10 - 15 mL soil saturation extract into a wide-mouthed porcelain crucible or a 150-mL Erlenmeyer flask.
2. Add 1 drop phenolphthalein indicator. If pink color develops, add 0.01 N H_2S_4O by a burette, drop by drop, until the color disappears.
3. Take the reading, y.
4. Continue the titration with 0.01 N H_2S_4O after adding 2 drops 0.1% methyl orange indicator until the color turns to orange.
5. Take the reading, t.
6. Always run two blanks containing all reagents but no soil, and treat them in exactly the same way as the samples. Subtract the blank titration reading from the readings for all samples.

Calculations

$$CO_3\ (\text{meq/L}) = \frac{2y \times N \times R \times 1000}{Wt}$$

$$HCO_3\ (\text{meq/L}) = \frac{(t - 2y) \times N \times R \times 1000}{Wt}$$

For Carbonate and Bicarbonate in soil:

Where: R = Ratio between total volume of the extract and extract volume used for titration.

N = Normality of H_2SO_4 solution.

Wt = Weight of air-dry soil (g)

10.8 Micro-nutrient Analysis by Wet Digestion

Full recovery of micro-nutrient cations (Zn, Fe, Mn, Cu) in high-silica containing plant tissues (like wheat, barley, rice, sugarcane, etc.) is not possible by dry ashing procedure. Therefore, this kind of plant materials should be wet-digested using HNO_3-$HClO_4$. The digestion procedure is adapted from Rashid (1986). Many other elements (like P, K, Ca, Mg, Na) can also be determined in the same digest.

Apparatus

Block-digester.

Vortex tube stirrer.

Atomic absorption spectrophotometer.

Flame photophotometer.

Reagent

Nitric Acid-Perchloric Acid (HNO_3-$HClO_4$), 2:1 ratio

To 1 L concentrated nitric acid add 500 mL concentrated perchloric acid.

Procedure

A. Digestion

1. Weigh 1 g dry plant material, and then transfer quantitatively into a 100-mL Pyrex digestion tube.
2. Add 10 mL 2:1 nitric-perchloric acid mixture, and allow to stand overnight or until the vigorous reaction phase is over.
3. Place small, short-stemmed funnels in the mouth of the tubes to reflux acid.
4. After the preliminary digestion, place the tubes in a cold block-digester, and then raise temperature to 150°C for 1 hour.
5. Place the U-shaped glass rods under each funnel to permit exit of volatile vapors.
6. Increase temperature slowly until all traces of nitric acid disappear, and then remove U shaped glass rods.
7. Raise temperature to 235°C.
8. Note time, when dense white fumes of perchloric acid appear in the tubes, and continue digestion for 30 minutes more.
9. Lift the tubes rack out of the block-digester, allow to cool a few minutes, and add a few drops DI water carefully through the funnel.
10. After vapors condense, add DI water in small increments for washing down walls of tubes and funnels.
11. Bring to volume with DI water. Mix the solution of each tube and then leave undisturbed for a few hours.
12. Each batch of samples for digestion should contain at least one reagent blank (no plant material).

B. Measurement

Decant the supernatant liquid and analyze Zn, Fe, Mn, Cu, Ca, and Mg in the aliquots by Atomic Absorption Spectrophotometry. Determine K and Na by Flame Photophotometry.

Calculations

For Micro-nutrient Cations in plant:

$$\text{Zn, Fe, Cu or Mn}\ (\text{ppm}) = (\text{ppm in extract - blank}) \times \frac{A}{Wt}$$

For Alkaline Earth Cations in plant:

$$\text{Ca, Mg, Na or K (ppm)} = (\text{ppm in extract - blank}) \times \frac{A}{Wt}$$

Where: A = Total volume of the extract (mL)

W = Weight of dry plant (g)

10.8.1 Determination of Ferrous

As total iron content in plant tissue does not indicate Fe nutritional status of plants, determination of ferrous iron (Fe^{++}) in fresh tissue by o-phenathroline extraction (Katyal and Sharma, 1980) is needed for the purpose. Then, ferrous content in the extracts can be measured by colorimetry or atomic absorption spectrophotometry.

Ferrous Extraction with o-phenanthroline

Apparatus

Spectrophotometer or colorimeter, 510-nm wavelength

Atomic absorption spectrophotometer

Reagents

Extraction Solution ($C_{12}H_8N_2$), 1.5% in HCl-buffer with pH 3.0 Add 15 g 1 - 10 o-phenathroline to about 850 mL DI water. Dropwise, add 1 N hydrochloric acid by continuously stirring solution until last traces of 1 – 10 o-phenathroline are solubilized. Final pH of the solution will be around 3.0. Make volume to 1-L volume with DI water.

B. Standard Stock Solution

Prepare working solution standards of iron containing 0, 1.0, 1.5, 2.0, 2.5, and 3.0 ppm Fe^{++} in the extraction solution.

Procedure

A. Extraction

1. Use carefully washed fresh plant tissues for ferrous analysis.
2. Weigh 2 g fresh (chopped with a stainless scissors) plant material into a 50- mL Erlenmeyer flask.
3. Add 20 ml extraction solution, and stir gently to ensure that all the plant tissue is completely dipped in the solution.
4. Close the flask using parafilm, and allow to stand for about 16 hours at room temperature.
5. Filter the contents through Whatman No. 1 filter paper.

B. Measurement

1. Ferrous content in the filtrate is determined by a Colorimeter at 510-nm wavelength or by an Atomic Absorption Spectrophotometer. Standards for Fe are run along with the plant extracts. 2. Ferrous content in plant tissue is expressed on oven dry weight basis, after determining moisture content in a sub-sample of fresh plant tissue.

Calculation

For Ferrous Iron in fresh plant tissue:

$$Fe^{++}\ (ppm) = ppm\ Fe^{++}\ (from\ calibration\ curve) \times \frac{A}{Wt}$$

Where: A = Total volume of the extract (ml).

Wt = Weight of oven-dry plant material (g)

Determination of Iron, Manganese, Zink and Copper

Iron, Manganese, Zink and Copper can be determined with the help of Absorption spectrophotometer (AAS)

Principle

The AAS is based on the principle that atoms of metallic element (Fe, Mn, Cu, Zn etc) which normally remain in ground state under flame conditions absorb energy which subject to radiations of specific wavelength. The absorption of radiation is proportional to the concentration of atoms of that element. The absorption of radiation by the atoms is independent of the wavelength of absorption and temperature of the atoms. These two features provide AAS a distinct advantage over flame emission spectroscopy. It also has greater sensitivity and accuracy.

Methods

A double beam atomic absorption spectrophotometer is required. The two most common oxidant/ fuel combinations used in atomic spectrometry are air-acetylene and nitrous oxide acetylene. Other flames that can be used are air- hydrogen and argon-hydrogen entrained air.

- Air acetylene: air acetylene is the preferred flame for the determination of approximately 35 elements by atomic absorption. The temperature ranges from 2125°C to 2400°C.
- Nitrous oxide acetylene: the nitrous oxide-acetylene flame has a temperature ranging from 2600°C to 2800°C and is used for the determination of elements which from refractory oxides. It is also used to overcome chemical interferances that may be present in flames of lower temperature. However, light emission from the nitrous oxide- acetylene flame is very strong at certain wavelengths. This may cause fluctuations in the analytical results for the element of interest is weak. Only the nitrous oxide burner head can be used with the nitrous oxide-acetylene flame.

10.8.2 Determination of Iron

Sample preparation: 1 g of oven dried plant samples are digested using diacid mixture and made upto 100 ml using deionized water on cooling.

Stock Standard Solution

1. Dissolve 1.00 g of pure iron wire in 50 ml of (1+1) analytical grade HNO_3. Dilute to 1 liter wit DI water.
2. Dissolve 7.022 g of analytical grade $(NH_4)_2$ Fe (SO_4)2 $6H_2O$ in 400 ml off DI water.

Add 5 ml of conc. H_2SO_4 and make up the volume with DI water. This would give a stock solution of 1000 ppm Fe. From this solution of 100 ppm is prepared which will be used for the preparation of final standard solution.

$$\text{Fe in ppm} = R \times \frac{100}{\text{sample}}$$

Similar calculation is fallowed fro Mn, Zn and Cu.

10.8.3 Dtermination of Manganse

The procedure is similar to as that of iron estimation.

Stock Standard Solution

1. Dissolve 1 g of manganese metal in a minimum volume of (1+1) NHO_3. Dilute to 1 liter with 1% (v/v) HCL.
2. Dissolve 3.076 g of $MnSO_4.H_2O$ In deionized water and make volume upto 1000ml .this will give a solution of 1000 ppm Mn. 10 m, of this solution is diluted to 100 ml to get a solution of 100 ppm Mn. this solution is used for preparation of final standards.

10.8.4 Determination of Zinc

The procedure is similar to that for iron estimation.

Stock Standard Stock

1. Dissolve 0.500 g of zinc metal in a minimum volume of (1+1) HCl and dilute to 1 liter with 1% (v/v) HCl.
2. Dissolve 4.398 g of $ZnSO_4.7H_2O$ in distilled water and volume is made upto 1000 ml this gives a solution of 1000 ppm zinc. 10 ml of this solution if diluted to 100 ml to get 100 ppm Zn solution. The final standard solutions are prepared from this 100 ppm solution.

10.8.5 Determination of Copper

The procedure is similar to that for iron estimation.

Stock Solution of Cu (1000 mg/I or ppm)

Dissolve 1 g of copper metal in minimum volume of (1+1) HNO_3. Dilute to 1 liter of (v/v) HNO_3.

Dissolve 3.929g $CuSO_4$. $5H_2O$ in distilled water and volume is made upto 1000 ml this gives a solution of 1000 ppm Cu. 10 ml of this solution is diluted to 100 ml to get 100 ppm solution of Cu.

Specification	Fe	Mn	Zn	Cu
Lamp current (mA)	30	20	6	15
Wave length (A°)	2483	2795	2139	3248
Linear range (ppm)	0-5	0-2	0-1.0	0-5
Slit width	2	2	7	7
Integration time (Sec)	2.0	2.0	2.0	2.0

For determination of Zn, Instead of hallow cathod lamps, Electrodeless Discharge Lamp

(EDL) is used. EDL will provide improved sensitive and lower detection limits than cathode lamp.

10.8.6 Determination of Molybdenum

Principle

In the thiocyanate method, an amber colored Mo complex of thiocynate is developed which is extracted into an organic solvent and the concentration determined calorimetrically.

Reagent

1. Hydrochloric acid: dilute as needed
2. Stannous chloride: ($SnCl_2$): Dissolve 20 g Sn Cl_2 in 20 ml of conc. hydrochloric acid (HCl) by heating, make to 200 ml with deionized water. Prepare fresh daily.
3. Ferric chloride hexahydrate ($FeCl_3.6H_2O$). Dissolve 49 g of $FeCl_3$. $6H_2O$ in deionized water and make to 1 liter.
4. Sodium nitrate ($NaNO_3$): Dissolve 42.5 g of $NaNO_3$ in deionized water and make to 100 ml.
5. Ammonium thiocyanate (NH_4SCN): dissolve 50 g of NH_4SCN in deionized water and make to 500 ml
6. 1-1 Di-isopropyl ether (the ether is washed before use with a mixture of one third $SnCl_2$, one third HN_4SCN and one third distilled water to remove impurities, if any. Place reagent grade 1-1 Di-isopropyl ether in a sepratory funnel and add wash solution equal to one tenth of volume of ether take. Shake thoroughly and let the organic phase separate from aqueous phase. Drain and discard the aqueous phase. Wash with 2 N HCl, again taking one tenth of volume of the ether. Shake and drain, repeat four to five times.

Standard Solution and Standard Curve

Standard Mo solution

Dissolve 0.150 g reagent- grade molybdenum trioxide (MoO_3) in 10 ml 0.1 N sodium hydroxide (NaOH), make slightly acidic with hydrochloric acid and make a 1 liter final volume with deionized water. Take 10 ml forms this solution and diluted to 1 liter for working standard having 1 μg of Mo / ml.

To prepare a working standard curve, add 5 ml of 6N HCl to a separating funnel. Add Mo standard solution. Add each separating funnel to a volume of 25 ml deionized water. The resulting solution is approximately 1.5 N with respect to HCl and approximates the normality of the unknown.

Procedure

Transfer the solution of unknown having about 1 to 3 of Mo to separtory funnel. Add in sequence 1 ml of $FeCl_3$ (to standard and blank only), 1 ml of $NaNO_3$, 5 ml of NH_4 SCN and $SnCl_2$. After the addition of each solution, shake remove and rinse the ground- glass stopper and the neck of the separatory funnel with deionized water using a fine tipped wash bottle. It is essential to keep the neck of the separatoryfunnel free of reagents because the final transfer of the Mo complex in the ether phase will be made by pouring it from the neck of the funnel. This is done by eliminate the possible transfer of residue reagents and moisture in the stem and stop cock if the ether phases were drawn off. Finally add 5 ml of the washed isopropyl ether and shake.

Allow the ether phase to separate, remove the ground glass stopper, rinse and wipe off the neck of funnel with tissue paper and draw off the aqueous phase. Transfer the ether phase to 15 ml centrifuge tube by pouring from top of the separatory funnel. Measure the transmittance at 475 nm on the colorimeter and estimate Mo in the sample using standard curve.

References

Baethgan W E, Alley M M. (1989) Plant anal. 20, 991 – 969. A manual colorimetric method of measuring ammonium N in soil and plant Kjeldahl digest. Comm. Soil Sci.

http://pubs.rsc.org/en/content/articlelanding/1964/an/an9648900276#!divAbstract

http://www.tandfonline.com/doi/abs/10.1080/00103629909370348

https://www.epa.gov/sites/production/files/2015-08/documents/method_351-1_1978.pdf

https://www.ncbi.nlm.nih.gov/pmc/articles/PMC440027/pdf/plntphys00342-0135.pdf

Tandon HLS (Ed.) 2004. Methods of analysis of soils, plants, water and fertilizer, Development and Consultation Organisation, New Delhi. India. pp. 144+ vi.

11
Chapter

Microscopic Techniques for Plant Tissue Analysis

11.1 Introduction

Plant anatomy plays an important role in the understanding of plant biology. A realistic interpretation of morphology, physiology and phylogeny must be based on a thorough knowledge of the structure of cells and tissues. Furthermore, the knowledge of the plant structure is also essential to solve many important everyday problems such as the identification of unknowns, food contaminants, and forensic problems. The aim of the laboratory exercise is to introduce the students to some basic anatomical organization of the plant organs, as well as cell and tissue characteristics.

For the study of plant tissue, Light microscopy as a matter of principle requires the passage of light through the object of study into the optical system of the microscope. An alternative to sectioning methods, which are by nature quite arduous, is provided by a variety of clearing techniques. These techniques, whereby thick masses of tissue are made translucent through specific chemical treatment, fall into two categories.

In the first category, clearing is accomplished by removal from the cells much of the protoplasmic content which tends to make the tissue opaque. Study can be accorded only to the cleared portion that remains, and methods of this kind are, therefore, restrictive.

In the second category by contrast, the tissue is structurally unaltered by the treatment process, and yet it becomes uniformly translucent. Although the mechanism operative is not presently understood, it is apparent that cellular organelles which ordinarily differ markedly in their refractive properties become more closely uniform in this regard.

The effect of refractive uniformity in tissue examined with ordinary bright-field optics is a very low contrast among the structural components of the cells. However, with phase contrast or Nomarski interference optics, which intensifies the small refractive differences among organelles, cellular structure can be examined throughout thick pieces of cleared tissue. Through selective staining of specific types of cells or cell organelles, techniques of this second category can be used effectively in conjunction with bright-field optics. Themethods in the second category, the 4½ clearing technique has been rather broadly applied in the study of plant tissue. The following four exercises are designed to acquaint biology students with some of the underlying principles of this technique, with some of its specific uses, and with its limitations.

11.2 Objectives

1. To study the processing of the tissue and preparation of blocks for microtome sectioning.
2. To study the sections of the plant tissue and examine the sections with the light microscope.
3. To study the distinguishing feature the cells that compose the tissue of rachis, stems, roots, and leaves.
4. To study the histochemistry of plant tissue
5. To study the cell damage during tissue development process.

11.3 Review of Plant Tissue Systems

It takes practice to make fine and good sections in which we can distinguish the feature of various cells. Tissue is group of cells organized into a structural and functional unit. Based on position and functions of tissue Sach(1875) considered three basic systems in plants, viz.

The plant system consists of following tissues:

1. **Dermal tissue**
   a) Skin
   b) Single layer of tightly packed cells that covers and protect plants.
2. **Vascular tissue**
   a) Transport elements
   b) Xylem and phloem.
3. **Ground tissue:**
   a) Storage ,
   b) Photosynthetic
   c) Bulk of plant tissue.

11.3.1 Dermal Tissue

The dermal tissue system, which covers the entire plant surface, protects the plant and regulates the flow materials between the plant and its environment. It is composed of various components such as epidermal cells, stomata, trachoma's, glands, and modified special structures.

1. Epidermis is an outer protective covering tissue of plant roots, leaves, and stems of non-woody plants. It contains closely packed epidermal cells.
2. Waxy cuticlecovers the walls of epidermal cells, minimizing water loss and protecting againstbacteria.

3. In roots, certain epidermal cells are modified into root hairs that increase surface area of the root for absorption of water and minerals and help to anchor plants in the soil.
4. Different protective hairs are produced by epidermal cells of stems and leaves.
5. Epidermal cells are modified as glands to secrete protective substances.
6. On the lower epidermis of dicot leaves, and both surfaces of monocot leaves, special guard cells form microscopic pores (stomata) and regulate gas exchange and water loss.
7. In older woody plants, the epidermis of the stem is replaced by cork tissue which is of bark.
    a) Cork is outer covering of the bark of trees; composed of dead cork cells that may besloughed off.
    b) Cork cambium is lateral meristem that produces new cork cells.
    c) As cork cells mature, they encrust with the lipid suberinthat renders them waterproof and inert.
    d) Cork protects a plant and makes it resistant to attack by fungi, bacteria, and animals

11.3.2 Vascular Tissue

The vascular tissue system: which consists of an arrangement of veins, transport of water and nutrients from root to leave and sugar to roots within the plant.

a) Xylem passively conducts water and mineral solutes upward through a plant from roots to leaves.
b) Xylem contains **tracheids** and **vessel** elements.

a) Tracheids

1. Tracheids are smaller, hollow, thin, long nonliving cells with tapered overlapping ends.
2. Water moves across end and sidewalls because of pits or depressions in secondary cell wall.

b) Vessel Elements

1. Vessel elements are hollow non-living cells lacking tapered ends.
2. They are larger than tracheids.
3. They lack transverse end walls.
4. They form a continuous pipeline for water and mineral transport.

c) Sclerenchyma

1. Xylem also contains Sclerenchyma cells to add support.

b) Vascular rays

1. It is flat ribbons of **parenchyma** cells between rows of tracheids; they conduct water and minerals across the width of the plant.

2. Phloem

The Vascular tissue that conducts the organic solutes in plants, from the leaves to the roots; it contains **sieve-tube cells** and **companion cells** is called phloem.

a) Sieve-tube Cells

1. Sieve-tube cells contain cytoplasm but no nucleus.
2. They are arranged end to end.
3. They have channels in their end walls (thus, the name "sieve-tube"), through which plasmodesmata extend from one cell to another.

b) Companion Cells

Companion cells are closely connected to sieve-tube cells by numerous plasmodesmata.

They are smaller and more generalized than sieve-tube cells.

They have a nucleus which may control and maintain the function of both cells.

They are also thought to be involved in the transport function of phloem.

Vascular tissue extends from root to leaves as vascular cylinder (roots), vascular bundles (stem) and leaf veins.

11.3.3 Ground Tissue System

The ground tissue system fills the space between the dermal and vascular tissue and serve a variety of functions including support, photosynthesis and storage.

a) **Parenchyma** are the least specialized of all plant cell types.

1. Cells of this type contain plastids (e.g., chloroplasts or colorless storage plastids).
2. They are found in all organs of a plant.
3. They divide to form more specialized cells (e.g., roots develop from stem cuttings in water).

b) **Collenchyma** resemble parenchyma but has thicker primary cell walls.

1. Collenchyma cells are uneven in the corners.
2. They usually occur as bundles of cells just beneath epidermis.
3. They give flexible support to immature regions of plants (e.g., a celery stalk is mostly collenchyma).

c) **Sclerenchyma** cells have thick secondary cell walls.

1. They are impregnated with **lignin** that makes the walls tough and hard.
2. They provide strong support to mature regions of plants.
   i) Most cells of this type are nonliving.
   ii) Sclerenchyma cells form fibers (used in linen and rope) and shorter sclereids (found in seed coats, nut shells, and gritty pears).

11.4 Histological Techniques for Plant Tissue Analysis

Methods of histology and histo-chemistry have a common origin and common core of similar procedures. These involve tissue fixation, dehydration, paraffin infiltration, and embedding, sectioning and staining. The preparations obtained by these methods are very informative and objectives of these have research and industrial application.

11.4.1. Sectioning of Plant Material

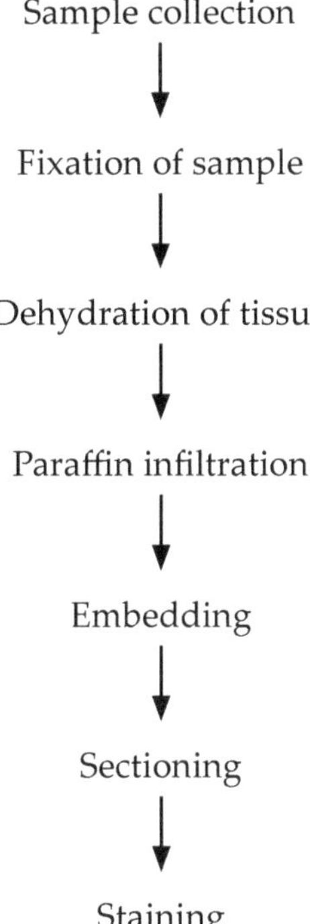

Sample Collection

Select the tissues at right stage of development, cut the materials with least possible injury. Cut the sample as possible as thin (approx 1.2 cm diameters).

11.4.1.2 Fixation of Plant Tissue

Fixing is the preservation of all cellular and structural elements in as nearly the natural living condition as possible.

11.4.1.3 Classification of Fixatives

a) **Acid fixation**

It preserves image particularly well the chromosomes, nucleoli, and spindle mechanism. Nucleoplasm and mitochondria are dissolved.

b) **Basic fixatio:**

Itpreserve fluids image mitochondria, nucleoplasm, and in some instances nucleoli and vacuoles. Chromatin and the spindle mechanism are dissolved.

c) **Coagulant fixatives**

The protoplast to form into a netlike structure and are the most numerous types. Some of the common coagulant fixatives are ethanol, methanol, chromium trioxide, picric acid, mercuric chloride and acetone.

d) **Non-coagulant fixatives**

It is exemplified by formaldehyde, glutaraldehyde, acrolein, osmium-tetroxide, acetic acid, and potassium dichromate. Some **non-coagulant** fixative can render tissue resistant to coagulation by coagulating substances during post-fixation and subsequent processes, thus permitting design of dual fixation systems.

e) **Additive fixatives**

It contains such as aldehydes crosslink with structures in the cell, namely proteins, and thus become a corporeal part of the cell structure. For some histo-chemical purposes such as the use of the Schiff reagent this may not be desirable.

Note

No single substance has been found to meet the requirements for successful fixation of all specimens. Fixatives usually consist of several ingredients in such proportions that there is a balance between the respective shrinking and swelling actions of the ingredients.

11.5 Fixation of Plant Tissue

11.5.1 FAA Fixation of Plant Tissue

Although FAA is a very harsh fixative and not balanced to cellular pH or osmoticum, it penetrates rapidly. Also, plant tissues can be stored in FAA almost indefinitely. FAA (or FPA) is stable and has good hardening action, and most material may be stored in it for years. These properties make it suitable for large or impervious objects such as woody twigs and stems, tough herbaceous stems, and old roots. The high concentration of alcohol is likely to produce shrinkage of succulent materials and also precipitation of the proteins, which gives a coarse appearance to the cytoplasm. However, it is possible to develop a formula for some apparently tender subjects and even for filamentous algae. A balanced formula can be worked out by varying the acetic acid, which has a swelling action on protoplasm, from 2 to 6% by volume. The acetic acid also preserves nuclear structure. The formaldehyde and alcohol have a shrinking effect and should be held at the indicated concentrations. When making trial of variations from fundamental formulas, fix a trial lot in the formula to be tested, and a check lot in the standard formula. Process the two lots simultaneously and identically and compare the cellular detail in both lots.

Materials and Solutions

1. FAA (Formalin-Acetic-Alcohol) (100 ml)

- Ethyl alcohol ------------------------------ 50 ml

- ➢ Glacial acetic acid -------------------------- 5 ml
- ➢ Formaldehyde (37-40%) ------------------- 10 ml
- ➢ Distilled H2O ------------------------------- 35 ml

2. FPA (Formalin-Propionic-Alcohol) (100 ml)

- ➢ Ethyl alcohol ------------------------------- 50 ml
- ➢ Propionic acid ----------------------------- 5 ml
- ➢ Formaldehyde (37-40%) ------------------- 10 ml
- ➢ Distilled H2O ------------------------------- 35 ml

Procedure

1. Fix pieces of thin leaf for 6-12 hours.
2. Longer fixation may make sectioning difficult because the vascular tissue may become much harder than the protoplasmic constituents.
3. Thick leaves or pieces of small stem require at least 24 hours. Woody twigs should be kept in FAA at least a week before continuing the processing for embedding. Materials do not need to be washed after FAA. The ingredients of this fluid are soluble in the dehydrating agents and are thus removed before infiltration is begun.
4. An extensively used formula consists of FAA containing mercuric chloride ($HgCl_2$) to saturation (1 g/10 ml FAA). This fluid penetrates and hardens tissues rapidly. It preserves bacterial zoogloea in plant tissues, thus being useful in pathological studies.

Note

The alcohol may be increased to 70%. Prolonged storage in fluids containing mercuric chloride is undesirable. The tissues should be transferred after 48 hours, or at most a week, to a fresh solution of the original formula which does not contain mercuric chloride. After four or five changes of the latter solution, most tissues may be stored indefinitely in the last change. Mercuric chloride is a highly poisonous substance, and fixatives containing it should be mixed and used in a hood.

11.5.2 Glutaraldehyde Fixation Solution (10 ml)

Glutaraldehyde-Osmium fixation is the most popular procedure for electron microscopy and also yields excellent preparations for the light microscope.

Materials and Methods

- ➢ 3% Glutaraldehyde ------------------------------- 0.6 ml of 50% Glutaraldehyde
- ➢ 0.02 M Sodium Phosphate Buffer (pH 7.0-7.4) --- 0.2 ml of 1 M Sodium Phosphate Buffer
- ➢ Distilled H_2O ------------------------------------ 9.2 ml

11.5.3 Osmium Fixation Solution (10 ml)

- ➢ 1% OSO_4 -------------------------------------- 0.1 g

- 0.02 M Sodium Phosphate Buffer (pH 7.0-7.4) --- 0.2 ml of 1 M Sodium Phosphate Buffer
- Distilled H_2O ------------------------------------ 9.7 ml

Procedure

1. Cut the tissue into very fine pieces.
2. For electron microscopy the pieces should be very fine indeed-about the size of a pinhead.
3. A convenient way to do this is to use sheets of dental wax.
4. Dissect the pieces on a pool of Glutaraldehyde Fixation Solution placed directly on the wax.
5. Use a clean sharp razor blade. Spray acetone on the blade and wipe off with Kim wipe to clean the blade.
6. Make in the hood. After mincing the tissue, use the blade as a scoop and place the tissue in a vial of Glutaraldehyde fixation solution at 4°C.
7. Change the Glutaraldehyde Fixation Solution after 2-3 hours and allow fixing for another 12-16 hours in the refrigerator.
8. Wash the tissue in cold 0.02 M Sodium Phosphate Buffer (pH 7.0-7.4) by making several changes at 30 min. intervals.
9. If the glutaraldehyde is not removed it will reduce the osmium which is added at the next step, and the reduced osmium does not have good fixative action. In addition the tissues tend to become coated with reaction product, and this decreases their permeability.
10. The tissue is postfixed in Osmium Fixation Solution for 1-2 hours at 4°C.
11. The osmium should be added in the hood; after use the waste osmium should be collected in a large screw-capped waste jar which should also be kept in the hood. Pasteur pipettes in conjunction with gum rubber bulbs are useful for transferring solutions.
12. Wash out the osmium with three changes of dehydrating fluids (ethyl alcohol and acetone) at 10-30 minutes each at 4°C.
13. Glutaraldehyde has rather slow penetration and may not suffice by itself for many subjects. The rate of fixation can be improved by adding formaldehyde and/or acrolein to the glutaraldehyde. A useful general procedure is as follows.

Place 1.0 g para-formaldehyde in 50 ml of H_2O at 60-70°C.

↓

Clear the solution withNaOH.

↓

Cool the solution to room temperature.

14. Add 6 ml of 50% glutaraldehyde and dilute the whole solution to 100 ml with 0.02 M Sodium Phosphate Buffer (pH 7-7.4).
15. Fix the tissue in the above solution for 2-6 hours at 4°C.
16. Change to Glutaraldehyde Fixation Solution for 12-24 hours.
17. Wash tissue and postfix in Osmium Fixation Solution at 4°C.

11.5.4 Glutaraldehyde-Paraformaldehyde Fixation Solution (10 ml)

Materials and Methods

- 0.25% Glutaraldehyde ---------------------- 0.1 ml of 25% Glutaraldehyde
- 4% Para formaldehyde ---------------------- 1.1 ml of 37% Para formaldehyde
- 0.1 M Na-Phosphate Buffer (pH 7.2) ------ 1 ml of 1 M Na-Phosphate Buffer
- DEPC-treated H_2O ----------------------- 7.8 ml.

Procedure

1. For fixation, the material should be of a manageable size.
2. Smaller pieces of tissue, allow greater penetration of the fixative, dehydrating solutions, and embedding material.
3. After subdividing, place the material as well as a label stating the identity for the tissue in small vials with 10-20 times the volume of Glutaraldehyde-Paraformaldehye Fixation Solution per volume of tissue.
4. Remove air bubbles, which interfere with proper fixation, by placing the unstoppered vials in an aspirator or vacuum desiccator under vacuum.
5. Aspirate for short intervals of time until material sinks to the bottom or just under the surface.
6. Meristematic tissue should be aspirated for a very short interval.
7. Tap the specimen bottle gently to aid in the removal of air bubbles.
8. Very buoyant material should be placed in a tall thin vial and held under the surface with a cheesecloth plug.
9. Most pieces are submerged after vacuuming; usually any floating piece will sink if pushed under the surface.
10. Be sure to aspirate the same type of tissue together; *i.e.*, do not mix highly vacuolate material with meristematic tissue.
11. If a specimen is too large or especially recalcitrant, leave the tissue overnight under vacuum, making certain that enough liquid is present in the vial to cover the tissue.
12. Fix the tissue at 4°C for 2-3 hours depending on the size of the tissue.
13. Rinse fixed tissue in 0.1 M Na-Phosphate Buffer (pH 7.2) for 30 min. Repeat.

11.6. Dehydration of Plant Tissue

Dehydration is the process of removing water from the fixed tissue and hardened tissues. Since rapid dehydration causes cells and tissues to become a distorted, the process must be slow. Generally two methods are in use for this purpose

1. Dehydration of tissues in non solvent of paraffin and then are transferred to a paraffin solvent (which also act as a clearing agent)
2. Dehydration of tissues in a solvent of paraffin.

Step	Ethanol	Time
1	5%	30 min
2	25%	30 min
3	50%	30 min

11.7. Paraffin Infiltration and Sectioning

After fixing, the water must be removed from the tissue before it is embedded in a water-insoluble matrix such as paraffin. Dehydration is accomplished by exposing the tissue to higher concentrations of ethanol or acetone. Tertiary Butyl Alcohol (TBA) is routinely used for paraffin material because paraffin is miscible with TBA.

From the 50% ethanol, dehydrate the tissue according to the TBA dehydration series as follow.

Step	100% Ethanol	TBA	Distilled H_2O	Time
A	50 ml	10 ml	40 ml	30-60 min
B	50 ml	20 ml	30 ml	30-60 min
C	50 ml	35 ml	15 ml	30-60 min
D	40 ml	55 ml	5 ml	30-60 min
E	25 ml	75 ml	------	30-60 min
F	------	100 ml	------	Overnight

- TBA solidifies at room temperature (25.5°C), so place the vials with absolute TBA on top of an oven or in another warm place.
- Follow with three changes of absolute TBA in Step F; keep the tissue in one change of absolute TBA overnight.
- Material should now be transferred to aluminium weighing dishes and filled 1:1 (v/v) with 100% TBA and molten paraffin.
- The paraffin should be filtered to make certain that it is particle-free.
- Place the aluminium weighing dishes in a warm oven (60°C) overnight.
- As the TBA evaporates, the paraffin will gradually infiltrate the tissue.
- Pour off any remaining mixture, and add fresh melted paraffin.
- Make one change with additional paraffin, and let the paraffin infiltrate the tissue overnight. Make 2-3 more changes of molten paraffin (every 4 hours). The material is now ready to embed.
- Have tools warm (use an alcohol lamp).

- Quickly transfer the aluminium dishes to a warming table and orient the tissue with warmed needles.
- When oriented, gradually move the dishes to a cooler area of the warming table.
- Allow the paraffin to solidify and cool completely at room temperature before transferring the embedded tissue to 4°C.
- Store at 4°C until use.
- Once a trimmed specimen and knife have been mounted in the microtome, sectioning can begin. Position the specimen before putting the knife into place.
- The block face of the specimen can either have a rectangular or a trapezoidal form, but in either case, make sure that the block is exactly parallel to the knife edge. Thus, as each section is cut, it will press evenly upon the trailing edge of the previous section and detach it cleanly from the knife edge with a minimum of wrinkling. Check that there is no grease, dust, or other contaminants on the block face or knife edge, as these may damage the tissue.
- Move the wheel of the microtome with a steady, even stroke.Hold the camel's hair brush in one hand to lift ribbons of sections as they come off the knife.
- Spread the ribbons in an appropriate container, or place them directly onto the slides. Examine the sections under the dissecting microscope to determine which sections you wish to affix to a slide.
- Prior to affixing them to the slides, sections can be stored overnight or longer in the storage box at 4°C.
- To affix the sections to the slides, place several drops of DEPC-treated H_2O on the slide. Use positively charged slides, which are commercially available.
- With a brush, place the sections on the flooded area, and very gently pull on their edges with dissecting needles to align them on the slide.
- Place the slide on a warming table (40-50°C) until any wrinkles in the ribbon are removed.
- Remove excess liquid with Kimwipes, and dry the slides overnight on the warming table in a dust-free area.
- Transfer the slides to a slide box and keep them at 4°C until use.
- Slides can also be deparaffinised at this point by putting them in a slide rack and subjecting them to two changes of xylene (in a fume hood), 30 min each.
- Remove the slide, tray to paper towels.
- After the xylene has evaporated, the sections may be examined with the microscope to determine if they are suitable for hybridization.
- The slides can be stored at 4°C indefinitely.

11.8. Over Fixation

Poor hybridization may be caused by inaccessibility of tissue mRNA. To prevent this, decrease the fixation time.

11.9 Under Fixation

It can lead to diffusion of tissue mRNA as well as poor tissue preservation. The optimal time must be determined empirically for each tissue type. Loss of tissue sections during the hybridization steps may be due to poor quality slides. Use positively (+) charged slides, which are commercially available.

11.10. Sectioning of Plant Tissues

11.10.1 Free Hand Sectioning

Plant parts are too thick to be mounted intact and viewed with a microscope. In order to study the structural organizations of the plant body the sections have to be made so that enough light can be transmitted through the specimen to resolve cell structure under the microscope. A free hand section is the simplest method of preparing specimens for microscopic viewing. This method allows one to examine the specimen in a few minutes. It is also suitable for a variety of plant materials, such as soft herbaceous tissue. The fixation of materials is generally not required for temporary preparations. Patience, experience and perhaps inherent skill are the chief requirements for this technique.

11.10.2 Microtome

These are mechanical devices for cutting uniform sections of the tissue of appropriate thickness. All microtome's other those used for producing ultra thin sections for election microscopy depend upon the motion of a screw thread in order to advance the tissue block on knife at a regulated number of microns.

Motion of screws can be direct or through system of gears or leavers to magnify the movements.

1. Paraffin Sectioning

Materials

1. Microtome
2. Water bath preferably thermostatically controlled

   Thermostatically controlled for paraffin wax of melting point 56ºC, a water temperature of 45ºC is sufficient ordinary distilled water is satisfactory; addition of a trace of detergent to water is beneficial in flattening of sections
3. Hot plate or drying oven thermostatically controlled

   Drying of sections at around the melting point of wax is satisfactory
4. Fine pointed forceps, Small hair brush and seeker needed to remove folds and creases in sections after floating out
5. Scalpel

6. Clear cloth or paper towel
7. Slide rack

   Majority of sections fit comfortably on a 76 x 25 x 1.2 mm slide.
8. Section adhesive:

   An adhesive is a substance which can be smeared on to the slides so that the sections stick well to the slides. Albumin, Gelatin, Starch, Cellulose, Sodium silicate, Resin and Poly L Lysine are the adhesives used for the above purpose.
1. Ice cube
2. Marker pencil

   Needed to write the identification details like name or specific number

Section Cutting Paraffin Embedded Tissue

Fixing of Block

1. Fix the block in the block holder on the microtome knife in such as position that it will be clear of the knife when it is in position., block may be fixed directly or it may be fixed to a metal carrier which in turn is fixed to the microtome.
2. Insert the knife in the knife holder and screw it tightly in position.
3. Trimming of the tissue block by moving forward so that wax block is almost touching the knife. To trim away any surplus wax and to expose a suitable adjuster are set at 15 to 20 microns.
4. On exposing a suitable area of tissue the section thickness is set to the appropriate level for routing purposes to 4-6 microns.
5. Apply ice to the surface of the block for a few seconds and wipe the surface of block free of water. This step is optional but makes sections cut easily.
6. Note that the whole surface of the block will move parallel to the edge of the knife in order to ensure a straight ribbon of sections.
7. The microtome is now moved in an easy rhythm with right hand operating the microtome and left hand holding the sections away from the knife. The ribbon is formed due to the slight heat generated during cutting, which causes the edges of the sections to adhere. If difficulty is experienced in forming the ribbon it is sometimes overcome by-rubbing one of the edges of the block with finger.
8. During cutting the paraffin wax embedded sections become slightly compressed and creased. Before being attached to slides the creases must be removed and the section flattened. This is achieved by floating them on warm water. Thermostatically controlled water baths are now available with the inside coated black. These baths are controlled at a temperature 4-6ºC below the melting point of paraffinwax. It is easy to see creases if the inside of water bath is black.

9. The action in floating out must be smooth with the trailing end of ribbon making contact with water first to obtain flat sections with correct orientation, floating out with the shiny surface towards the water is essential. When the ribbon has come to rest on water the remaining wrinkles and folds are removed by teasing apart by using forceps or seeker.
10. ***Picking up sections*** – The ribbon of sections floating on water is split into individual or groups of sections by use of forceps or seekers. Picking up a section on slide is achieved by immersing the slide light lysmeared with adhesive vertically to three fourths of its length bringing the section in contact with the slide. On lifting the slide vertically from the water, the section will flatten on to the slide. The sections are then blotted lightly with moistened blotting paper to remove excess water and to increase contact between section and slide. For delicate tissues or when several ribbons of sections are placed on the slide,omit the blotting instead keep the slide in upright position for several minutes to drain.
11. ***Drying of section:*** Sections are then kept in incubator with a temperature 5-6ºC above the melting point of wax i.e. at 60ºC for 20-60 minutes. It is better to overheat than underheat. If the sections are not well dried they may come off during staining. The sections should not be allowed to dry without a good contact with the slide;such sections will come off during staining.

11.11. Staining Plant Tissue Sections

a) Safranin Staining Solution (10 ml)

Materials

- 2% Safranin O --------------------------- 0.2 g
- 95% Ethyl alcohol ------------------------ 10 ml

Procedure

Dry Paraffin Embeded Sections at 60°C Overnight.

1. Deparaffinize slides in xylene.
2. Cover the sections with several drops of Safranin Staining Solution.
3. Rinse with distilled H_2O and air-dry.
4. Mount with a suitable medium.

Results

B) Toluidine Blue Staining Solution (10 ml)

The main advantage of staining using toluidine blue is the short time, required for processing. The sectioned material can be stained without deparaffinization, thus eliminating the hydration-dehydration steps.

Materials

- 0.05% Toluidine Blue O --------------------------- 0.5 ml of 1% Toluidine Blue
- Distilled H_2O ------------------------------------- 9.5 ml{or, 0.1 M Citrate

Phosphate Buffer (pH$_4$-6) ------- 9.5 ml

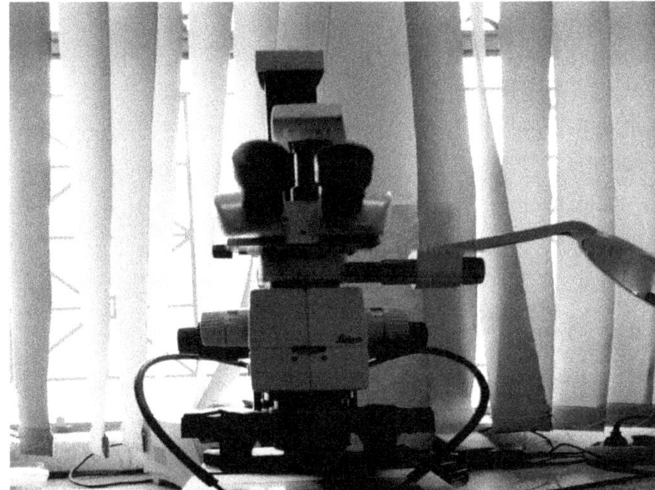

Procedure

1. Stain Non-deparaffinised slides in Toluidine Blue Staining Solution for 15-30 min.
2. Rinse slides in distilled water briefly and air-dry.
3. Remove paraffin with two changes of xylene overnight.
4. Mount slides with synthetic resin.

C) Hematoxylin & Eosin Staining Protocol

Note: Prior to staining, slides must be baked for at least two at a minimum temperature of 70°C to inhibit detachment of sections during the staining procedure.

Materials

- Ethanol
- Eosin-Phloxine
- Histoclear

Staining Procedure

1. Stain in Weigert's Hematoxylin, working solution, for 6 minutes.
2. Drain the slides.
3. Differentiate in 70% Acid-Ethanol, pH 2.5, two changes, and three quick dips each.
4. Wash in running tap water for 15 minutes. Drain excess water from slides.

5. Stain in working Eosin-Phloxine solution for 2 minutes.
6. Drain slides.
7. Differentiate and dehydrate in 95% ethanol, two changes, and 30 seconds each.
8. Completely dehydrate in 100% ethanol, three changes, and three quick dips each.
9. Clear in Histoclear, two changes, one minute each.
10. Fresh Histoclear and cover slip with mounting medium.

Results

Nuclei – deep blue-violet

Chromatin – blue-black

Nucleoli – varying shades of orange-pink to red-violet, depending on cell type.

Cytoplasm – varying shades of pink and orange

Collagen – pale pink.

D) Staining Section with Fast Green

This schedule has been found to be very satisfactory for particularly all types of anatomical material and it gives possible results with semi-cytological items. The use of tertiary butyl alcohol was first suggested by Johansen (1935). The method of making up the fast green solution was suggested by a somewhat similar procedure utilized by E.J. Kraus (described by Conant 1926) who however used gentian violet instead of fast green. The advantages of using the green are that it provides a greater contrast with safranin, and it does not fade so rapidly.

Staining Procedure

1. Deparaffinize slides in xylene for 5 minutes
2. Remove paraffin with two changes of xylene each for 5 minutes.
3. Rinse with Absolute alcohol for 2 minutes
4. Rinse for 5 – 6 times with 95 % alcohol.
5. Rinse for 5 – 6 times with 50% alcohol.
6. Stain with 1% safranin in 50 % alcohol for 6- 10 minutes
7. Wash with water
8. Remove the stain with 50% alcohol for 2-3 minutes.(in most cases destaining with acid alcohol is not necessary).
9. Rinse with 70% alcohol for 5-6 times
10. Rinse with 95% alcohol for 5-6 times.
11. Rinse with Absolute alcohol for 2 minutes
12. Rinse for 5-6 times with 40 % alcohol and 60 % xylol.
13. Fast green (prepared by adding a few pipettes of saturated clove oil solution of fast green FCF to a stenderof 40 % absolute alcohol and 60

% xylol. The solution should be fairly dark green, although the strength may vary considerable. If solution seems unstable or turbid, let it stand for a day or two and if it does not become clear add a pipette of absolute alcohol,) time to determine by trial. Try 2 or 3 minutes at fast. There is very little danger of over –staining, or of reducing the safranin.

14. Rinse with 40 % absolute alcohol and 60% xylol
15. After the xylene has evaporated, the sections may be examined with the microscope to determine if they are suitable for hybridization.
16. Mount with a suitable medium.
17. The slides can be stored at 4°C indefinitely.

Notes- A few section of each sample should be stained to check that the tissue is intact. This also provides a useful record of morphological details. (Often at the end of hybridization procedure tissue morphology is not as clear.)

Results

Nucleoli, Chromosomes, Cuticle and lignified cell walls stain red. The remaining organelles stain green.

11.12. Analysis of Tissue by Fluorescent Staining

11.12.1 Analysis of Xylem by Auto Fluorescence

Because of large content of lignin in cellular walls xylem shows auto fluorescence in UV part of spectrum and emits blue light. So, it is possible to observe them under fluorescence microscope without prior staining.

Materials

- Slides
- Cover slips
- Blades
- Water
- Dropper

Procedure

1. Cut relatively thin layer of stem or some other part of plant.
2. Place the section on microscopic slide.
3. Add few drops of water and cover with cover slip to avoid loss of liquid.
4. Observe under fluorescence microscope in UV light.

Observations

Mechanical cells characterized by lignin deposits in their cellular wall become visible. Lignins emit blue light in UV spectrum.

11.12.2 Analysis of Lignin Deposits using Eosin Staining

Staining cells with eosin allows differentiation among xylemic parts functioning as water transporter and those which are not. Eosin acts like xylem mobile dye and it moves easily through tracheas, membranes of pit, and parenchyma cells. Though, eosins stains lignin unselectively, solution of dye passes only across functional xylem vessels. Thus number of functional and non functional elements of xylem could be determined by comparison of yellow and blue fluorescence. In order to achieve this it is necessary to use fresh sample.

Materials

- Plant part
- Eosine dye solution
- Slides
- Cover slips
- Water

Procedure for Staining

1. Cut plant pieces and place into water, then by Gillette cut into 2-3 cm.
2. Cut ends submerged into dye solution. For successful staining it is recommended that upper end of cut parts to have leaves. Depending on length of stem and intensity of transpiration wait for 10 to 30 min. until dye reaches opposite end of stem (leaves turn red)
3. Wash excess of eosin from part of stem submerged.
4. Cut into 10 -20 μm wider sections. Place sections onto slide and put water drop.
5. Cover with cover slip and observe under light and fluorescent microscope leica DMLS (filter A 340-380 nm).

Observations

When observed under fluorescent microscope with UV filter, unstained cellular walls with deposits of lignin appear blue. While cell walls with lignin stained with eosin exhibit bright yellow fluorescence. Parenchymal cells which have no lignin in their walls are invisible after staining because they do not fluoresce.

11.12.3 Analysis of Phloem by Auto Fluorescence

Periderm tubes in Vitis vinifera possess several layers of phloem cells differentiated around mature strand of secondary phloem fiber within bark.

Materials

- Slides
- Coverslips
- Blades
- Water
- Dropper

Procedure

1. Fresh free hand, transverse sections were made along internodes of 1- 2 year old branches of grape cultivar.

2. Mounted the sections in water and put the cover slip.
3. Auto fluorescence was observed by viewing in Zeiss axiophotep ifluorescence microscope equipped with HBO 100 lamp and UV filter assembly (BP 395-490,FP 460, LP 470)

Observations

Subrised walls of phloem cells are bright yellow.

11.12.4. Analysis of Sieve Tubes

Aniline blue fluorescence is widely used in botanical histochemistry particularly to stain callose plugs in the phloem. Strong association of aniline blue fluorescence with β 1-3 glucans such as callose is attributed to lose packing of this polymer providing greater accessibility to them by fluorocrome.

Materials

- 0.01% aniline blue
- 0.05 M Na-K phosphate buffer
- Slides
- Coverslips

Procedure

1. For callose sections were stained with 0.01% aniline blue in 0.05 M Na- K phosphate buffer, pH 7.2 for 5 min.
2. Then mounted in buffer of same pH.
3. Sections were observed with epifuorescence with filter mentioned above.

Observations

Staining with aniline blue shows that secondary phloem is built of layers of sieve tubes and fibres. Callose in sieve plates exhibits bright light.

11.12.5 Analysis of Phenolic Compounds in Grape Leaves

Epidermal UV screening is assessed most frequently from absorbance of extracted phenolics. The distribution of phenolics in grape leaves can be investigated by fluorescence microscopy.

Materials

- Plant leaves section
- 0.5% (w/v) Ammonia
- Axioplan microscope

Procedure

1. Take cross sections of freshly harvested leaves of 70-100 μm thickness using hand microtome.
2. Samples examined in water.
3. Then immersed in 0.5% (w/v) ammonia kept on blotting paper.

4. Take fluorescence images using Axioplan microscope zeiss filter set no. 5 set that excites in range of 395-440 nm and detects fluorescence at wavelength >470 nm.

Observations

The presence of Phenolic compounds in upper epidermis indicated by ammonia enhanced green fluorescence.

11.12.6 Analysis of Fungal Infection

Fluorescent microscopy methods, fungal cells in sample are marked with fluorophore. When sample placed under microscope with specific filter labelled cells appear colored and more easily detected.

Materials

- Fresh plant tissue
- FAA fixed plant tissue
- 0.1% Aniline blue w/v
- 0.1 M KH_2PO4
- Slides
- Cover slips

Procedure

1. Take fresh plant tissue sections.
2. Stain the sections in 0.1 % Aniline blue prepared in 0.1 M KH_2PO_4
3. Fungal growth can be examined in Zeiss Axioscopmicroscope; equipped zeiss filter 6 and 18, neoflur objective, excitation filter BP 436/8 and BP 395-425.
4. Wax embedded FAA fixed plant tissue stained for 2 min in 0.5 aqueous Malachite green.
5. Then sections were washed in water to remove excessive stain.
6. Counter stain the sections with 0.01 % acridine orange prepared in phosphate buffer pH6, for 20 mins.
7. Wash in phosphate buffer.
8. Air dries the sections and mounts in immersion oil.
9. Observe the sections under fluorescence microscope with above set up.

Observations

With this setup conidia and fungal hyphae fluoresce yellow. Sections stained with acridine orange shows hyphae fluoresced in orange color.

11.12.7 Analysis of Chlorophyll by Autofluorescence

Plant tissues have high level of auto fluorescence in chloroplast and accumulate various secondary metabolites. In chloroplast endogenous auto fluorescence arises from variety of biomolecules including chlorophyll, carotene, Xanthophylls.

Chlorophyll the major contributor of auto-fluroscence has absorption band in blue region of visible spectrum which produces significant amount of fluorescence at wavelength > 600 nm when exited with wavelength between 420 nm and 460 nm.

11.12.8 Fluorescence Methods for Detection of Microorganism

Microscope based techniques are commonly used to enumerate microorganisms in environment without prior culturing. Fluorescence microscopy is often used to differentiate between viable and non-viable cells. The microbial cells in sample are labelled with fluorescent marker. Different fluorophores are available for cells of different types in different metabolic states. Fluorophore detects cell viability by tracking differences in cytoplasmic redox potential, electron transport chain activity, enzymatic activity, cell membrane potential and membrane integrity.

The most common dyes used are Acridine orange, DAPI, Ethidium bromide, Rhodamine.

Observations

- Acridine orange – Binds to both DNA and RNA, Single stranded nucleic acid emits orange red fluorescence and double stranded nucleic acid fluorescens green.
- DAPI- Fluorescens blue or bluish white when bound to DNA and yellow when bound to non DNA material.
- Ethidium bromide – Interacts with DNA then fluorescens red orange.
- Rhodamine- Protein specific fluorescent stain.

11.12.9 Detection of Intracellular Suberin

Intracellular suberisation is strongly associated with individual cells or cell layers displayed with fluid diffusion test. Test includes use of phloroglucinol + HCl and Sudan black to selectively quench auto fluorescence of lignin and suberin resp.

Blue violet excitation is used to enhance sudan test for suberin, cutin wax.

Procedure

1. Parrafin embedded tissues sectioned at 10 μm.
2. Stain with Sudan black for 5 min
3. Decolorize with 95 % ethanol.
4. Fluorescence was examined with fluorescence microscope with filter combination A (340-380 nm excitation and 415 nm barrier filter) and H_2 (390-490 nm excitation and 510-515 barrier filter)

11.12.10 Organelle staining- Endoplasmic Reticulum , Golgi and Mitochondria

Live plant cells of cell line will be stained with Dioc5 which stains endoplasmic reticulum,Ceramide which stains golgi apparatus, Mito tracker red which stains mitochondria.

- **Golgi apparatus**

The glycolipid ceramide associated with membranes sometimes it accumulates in golgi. Serve as convenient marker for organells. Ceramide itself is not fluorescent, so we will use it as indirect stain in which green fluorochrome BODIPY FL is conjugated to ceramide.

- **Endoplasmic Reticulum**

  ER Tracker localizes preferentially to membranes of endoplasmic reticulum staining them blue /white. The hydrophobic side chains of ER Tracker provide specificity to ER membranes. DiOC5 localizes preferentially to the membranes of the endoplasmic reticulum (ER), staining them green. The hydrophobic side chains of the lipophilic DiOC5molecule provide specificity for the ER membranes. This stain also will accumulate in the mitochondria to some degree. Pay attention to the fine network of staining rather than the large blobs. The fine network is the ER.

- **Mitochondria**

  MitoTracker Red is a vital stain that will be used to stain mitochondria. Sequestration of this red dye is dependent on a membrane potential; thus, the stain can also reveal information about the health of the mitochondria. The membrane potential is rapidly lost if the mitochondria are no longer able to generate an ion gradient, e.g. by damage to the membrane, inhibition of the electron transport system, etc.

Procedure

1. Cover slips with cells are arranged in a petridish.Wipe a micro slide clean, and label it.
2. Cut 3 inch square of parafilm, tape the parafilm flat on a bench and label with stain name.
3. Pipette 300 µl of stain onto parafilm.
4. Using a pair of forceps, remove cover slip from petridish with live cells making sure that cells are facing upward.
5. Dip the cover slip 2-3 times in PBS. Blot excess of PBS by touching the edge of cover slip with wipe.
6. Put the side of cover slip containing cell downward on drop of stain. Cover the slide with foil and stain cells for 15 min.
7. Rinse.
   1. Ceramide stained cells only- Set up fresh piece of parafilm with 300 µl PBS. After staining in ceramide for 15 minutes, rinse the cells in PBS by dipping coverslip in PBS 3-4 times.Then invert coverslip on fresh PBS for 15 minutes. Rinse cells in coverslip in beaker of PBS for 30 seconds. Drain out excess of PBS with wipe.
   2. Mitotracker red and DiOC5- After staining rinse the cells in PBS by dipping coverslip in PBS for 30 seconds.Briefly, touch the edge of coverslip with wipe to remove excess of PBS. Don't let coverslip dry out. Mount the coverslip onto drop of PBS on slide.

8. Put a drop of (20 μl) of PBS onto clean labelled slide. Using forcep place one edge of coverslip against slide and slowly drag it so that PBS spreads under cover slip. Try to avoid trapping air bubble as you lower cover slip. Using the wipe blot excess of PBS.
9. Seal the edges with narrow band of nail paint. The seal will prevent cells from drying out. The seal will prevent the cells from drying out while observing the cells.
10. When the nail polish is dry, rinse the cover slip with distilled water. This removes excess of PBS from slide. Dry the slide with wipe and be careful that not to press too hard on cover slip. Thoroughly dry the underside of slide.
11. Observe the cells. Start with 20 X objective using phase contrast. Change to 40 X without changing course focus knob. Adjust the focus using fine focusing.
12. To examine Golgi and ER turn the filter cube dial to NIB position. Open shutter by pushing in the shutter knob. Focus using fine focus knob.
13. Examine the mitochondria stain using WIG filter.
14. The DNA specific stain can be included so the cells will also stain for DNA. After examining organelles in interphase cells, switch to WU cube to examine Hoesht staining pattern. In interphase nuclei will appear as blue ovals.

11.13. Fluorescence Microscopy

The fluorescent stains reveal the structure within cell dramatically. In addition availability of different fluorocromes allows staining different structures simultaneously and observing distribution of structure relative to each other in same cell.

With regular light microscopy, when you look at sample, in addition to your sample you are observing background illuminating light. The light can make difficult to see weakly stained cellular components. With fluorescence microscopy, background light is eliminated so you see fluorescently stained object contrasted against black background. Fluorescent molecules absorb light of particular wavelength and then emit at longer wavelength. The fluorescent microscope takes advantage of this difference, filtering out the excitation light, so only the emitted light is seen. For example, the fluorochromeAlexa Fluor 488 is excited by blue light (maximally at 495 nm), yet it emits in the green (520 nm) range. To examine structures stained with Alexa Fluor 488, cells are illuminated with a blue light. The cells give off a green fluorescence; in addition, blue light is reflected back from the sample. However, dichroicfilters in the microscope block the blue light while allowing the green light to pass through. Then, as you look at the cells through the microscope, the Alexa Fluor 488 stained structures glow green against the black background. Fluorescence microscopy is frequently used in cell biology because of its increased sensitivity and because of the ability to stain specific parts of a cell.

However, there are not many fluorescent chemicals with innate high affinity for directly staining particular part of cell.

Stain class	Stain	Fluorochrome	Domain	Structure stained	Stain color
Direct	Mitotracker red	Mitotracker red	Mitotracker red	Mitochondria	Red
Direct	DiOC5	DiOC5	DiOC5	Endoplasmic reticulam	Green
Direct	DAPI	DAPI	DAPI	DNA (nuclei)	Blue
Direct	Hoechst 33342	Hoechst 33342	Hoechst 33342	DNA (nuclei)	Blue
Indirect	BIODEPY FL C5-ceramide	BIODEPY FL	Ceramide	Golgi Apparatus	Green
Indirect	Alexafluor 488 phalloidin	Alexafluor 488	Phalloidin	Actin	Green
Indirect	Texas red anti mouse 2ºantibody	Texas red	Anti mouse 2º antibody	Tubulin	Red

Hazards

- Wear gloves for all staining procedure. Many reagents are toxic. Avoid contact with skin, eyes, mouth.
- Glass cover slips are fragile. Try not to crack them and avoid impaling yourself with slivers. Clean up any glass shards immediately.
- Be careful while using razor blade with plants.
- Do not stare at the stray UV light from fluorescence microscope. Make sure that there is UV shield on your microscope. The shield is smoke colored piece of plastic in front of objective.

11.13.1 Toluidine blue O Staining

The stain toluidine blue O is excellent stain for histological study. TBO has advantage of being polychromatic dye i.e. it reacts with different chemical components of cells differently and result in multi colored specimen. The colors generated can provide information on nature of cell and its walls. TBO is cationic dye that binds to negatively charged groups. An aqueous solution of this solution is blue, but different colors are generated when dye binds with different anionic groups in the cell. For example, a pinkish purple color will appear when dye reacts with carboxylated polysaccharide such as pectic acid; Green, greenish blue or bright blue with polyphenolic substances such as lignin, tannins; and purplish or greenish blue with nucleic acid.

Stain Preparation

- Dissolve 0.1 g of toluidine blue O in 100 ml of benzoate buffer pH 4.4 (Benzoic acid 0.25 g, Sodium benzoate 0.29 g, water 200 ml).

This buffer is recommended for histochemical purposes. If benzoate buffer is not available, for general use tap water can be used as solvent for TBO.

Staining Procedures

1. Prepare sections by microtome and deparaffinise in xylene.
2. Flood the sections with aqueous solution of 0.1 % TBO solution for 1 min.
3. Gently remove the stain by using piece of filter paper. Wash the sections by flooding them with water followed by its removal. Repeat until there is no excess stain around the sections.
4. If required destain in alcohol by series of reducing concentrations and then in xylene.
5. Mount in DPX or in 30 % glycerol and put cover slip.
6. Remove excess mounting medium by gentle touching the edge of cover glass with filter paper.

Results

Pectin will be Red or reddish purple; lignin blue; other Phenolic compounds green to blue green. Thin walled parenchyma will be reddish purple; Collenchyma, reddish purple; lignified elements such as tracheary elements and sclerenchyma will appear green to blue green; Sieve tubes and companion cells, purple ; Middle lamella red to reddish purple; Callose and starch unstained.

11.13.2 Phlorogucinol HCl Test for Lignin

Lignin is commonly constituent in secondary wall of plant cells. Eg.Walls of xylem elements and sclerenchyma tissue. The Cinnamaldehyade end groups of lignin appear to react with phloroglucinolHCl to give red violet color. Although the reaction is not very sensitive, because of ease of staining, this procedure is still often used as one of the test for presence of lignin in plant cell wall.

Stain Preparation

- It is commonly prepared as a saturated solution of phloroglucinol in 20 % 2N hydrochloric acid. First dissolve phloroglucinol (2g) in 80 ml of 20 % ethanol solution and then add 20 ml of concentrated HCl (12 N) to it.

Staining Procedure

1. Prepare free hand sections or paraffin embedded sections.
2. Deparaffinise the sections in xylene.
3. Stain the sections in phloroglucinol HCl for 2 or more minutes if lignified cells are present specimen will turn red in few minutes.
4. Add a drop of 30 % glycerol or DPX .Place a cover slip over section before observing in microscope.
5. Examine the specimen at once hence color fades rapidly.

Results - Lignified wall becomes red.

11.13.3 - Iodine – Potassium – Iodide (IKI) Test for Starch

Iodine –potassium – iodide (IKI) is specific for starch. Apparently the basis of the reaction is accumulation of the iodine in the center of helical starch molecule. The length of starch molecule determines color of reaction. Shorter the molecule, more red color; longer the molecule more blue color.

Stain Preparation

- The IKI solution is prepared by first dissolving 2g of KI in 100 ml distilled water and adding 0.2 g of iodine in KI solution. Prepare this solution ahead of time as iodine takes some time to dissolve. Store the solution in dark glass bottle and cap tightly. Exposure to light and air degrades solutions usefulness. Iodine sublimates at room temperature.

Staining Starch Iodine- Potassium- iodine Test Procedure

1. Prepare the hand sections or paraffin embedded sections.
2. Place a drop of IKI directly on specimen.
3. Wait for few minutes and mount in 30% glycerol, Put the cover slip.
4. The specimen can be examined without removal of excess of IKI.

Results

Starch will give blue black color in few minutes. Newly formed starch may appear red purple.

4. Total Lipids

The mechanism of staining based on differential solubility. The sudan dyes are more soluble in apolar solvents. As a result they tend to dissolve more in structure such as the cuticle, lipid droplets, suberin which are hydrophobic.

Staining Solution

- Staining solution is prepared by dissolving 0.7 g of sudan IV in 100ml of propylene or ethelene glycol. Heat the solution to 100 ºc and stir it to several minutes. Filter the hot solution through whatman paper 2 No. Cool and filter again.

Staining Procedure

1. Prepare the sections as described above.
2. Stain the sections in sudan for about 5 min.
3. Transfer the sections to 85% propylene or ethylene glycol in water. And agitate the container gently for about30 sec This is to wash away excessive stain from sections which allows a better differentiation of stain in various sections.
4. Briefly rinse the sections in water and mount in glycerol.

Results

Fats, oils and waxes will stain red. The cuticle of leaves, suberized walls in cork cells and casparian cells if present will stain red because of lipidic nature of these structure.

11.13.4 Mounting of Sections

Histological sections which need to be examined for any length of time or tobe stored must be mounted under a cover-slip.

There are two types of mounting media:

1. Aqueous media - Used for material which is unstained, stained for fat,or metachroamtically stained.
2. Resinous media - For routine staining.

Aqueous Media

There are used for mounting sections from distilled water when the stainswould be decolorized or removed by alcohol and xylene, as would be thecase with most of fat stains (Sudan methods). Some stains, e.g. methylviolent, tend to diffuse into medium after mounting. This can be avoided byusing Highman's medium. Aqueous mountants require addition ofbacteriostatic agents such as phenol, crystal of thymol or sodium merthiolateto prevent the growth of fungi.

Glycerine is an a excellent routing mountant for fats stains.

Resinous Mounting Media

Natural or synthetic resins dissolved in benzene, toluene or xylene. Theseare purchased readymade. In case they become too viscous they may haveto be diluted with xylene. Following are some of these media.

1. Canada balsam - Natural resin (R.I. - 1.52)

   It is used as 60% resin by weight in xylene. H.&E stained slides arefairly well preserved but basic aniline dyes tend to fade and Prussianblue is slowly bleached. Slides take few months to dry.
2. D.P.X. (R.I. 1.52)

   Polystyrene resin dissolved in xylene as a 20% solution. It is mostcommonly used.

There are many other synthetic resins sold under various trade names e.g. Coverbond (R.I. 1.53), H.S.R. (Harlew synthetic Resin), Histoclad (R.I. - 1.54), Permount (R.I. 1.54), Pro-Texx (R.I. 1.495).

- Criteria of acceptable mounting media
  1. Refractive index should be as close as possible to that of glass i.e. 1.5.
  2. It should not cause stain to diffuse or fade.
  3. It should not crack or appear granular on setting.
  4. It should be dry to a nonsticky consistency and harden relatively quickly.
  5. It should not shrink back from edge of cover-glass.
  6. It should be free flowing and free from air bubbles.

11.15. Special Staining

11.15.1 Van Gieson Method

Staining of the connective tissue

Principle

In the routine staining method collagen, elastic fibres and smooth muscle appear pink or reddish in color. In the Van Gieson stain, collagen and most reticulin stain selectively with acid aniline dyes (acid fuchsin). Picric acid acts as counter stain for muscle and cytoplasm and form complex with the dyes. This complex has special affinity for collagen.

Materials

1. Solution A
   a) Haematoxylin 1.0 g
   b) Alcohol 95% 100 ml
2. Solution B
   a) 29% (w/v) ferric chloride 4 ml
   b) Conc. Hydrochloric acid 1.0 ml
   c) Distilled water 95.0 ml
3. Weight's iron heamatoxylin solution:
   a) Mix equal quantities of solution A and solution B.
   (Color of this reagent should appear violet black)
4. Van Gieson's solution
   a) Saturated aqueous picric acid – 10 ml
   b) 1% (m/v) acid fuchsin – 1.5ml
   (It should be freshly prepared)

Procedure

1. Deparaffinise with xylene
2. Hydration take sections to water
3. Stain with Weigert's haematoxylin for 20-40 minutes
4. Wash in distilled water
5. Van Gieson stain 1-3 min
6. Rinse well in distilled water
7. Dehydrate in absolute alcohol (2 changes)
8. Clear in xylene (2 changes)
9. Mount in DPX

Results

1. Collagen Red
2. Muscle and Cornified epithelium Yellow
3. Nuclei Blue to Black

11.15.2 Gomori's Method for Reticulum

Principle

In the connective tissue, reticulum appears as a fibrillary extra cellular framework. Reticular fibres have low natural affinity for silver salts and require pre-treatment with heavy metal solutions like ferric ammonium sulphate to enhance the selectivity of impregnation Silver in alkaline solution is in a state readily able to precipitate as metallic silver. Upon treatment with a reducing agent, silver taken up by the tissue in unreduced form is, converted to metallic silver which is deposited at the sensitised site.

Materials

1. Ammonical Silver Solution

To 10 ml of 10% silver nitrate solution add 2.5 ml of 10% aqueous solution of potassium hydroxide; add 28% ammonium hydroxide drop by drop while shaking the container continuously until the precipitate in completely dissolved.

Add again 4 drops of silver nitrate solution for every 10 ml of silver nitrate used. Make the solution with distilled water to twice its volume use acid clean glassware

2. 0.5% potassium permanganate

Potassium permanganate –0.5 g

Distilled water 100 ml

3. 2% potassium meta-bisulphite

Potassium meta-bisulphate 2.0 g

Distilled water 100 ml

4. 2% ferric ammonium sulfate solution

Ferric ammonium sulphate 2 g

Distilled water 100 ml

5. 20% formation solution

Formaldehyde 20 ml

Distilled water 80 ml

6. 0.2% gold chloride solution

Gold chloride solution 1% - 10 ml

Distilled water - 40 ml

7. 2% sodium thiosulphate solution

Sodium thiosulphate 2 g

Distilled water 100 ml

Procedure

1. Deparaffinze and hydrate to distilled water
2. Oxidize in potassium permagnate solution for 1 minute
3. Rinse well in tap water – 2 min
4. Differentiate with potassium metabisulphite solution for 1 minute.
5. Wash in tap water for 2 min
6. Sensitize in Ferric ammonium sulphate solution for 1 min.
7. Wash in tap water for 2 minutes follow with two changes of distilled water 30 seconds each.
8. Impregnate in the silver solution for 1 minute
9. Rinse in distilled water for 20 seconds
10. Reduce in formalin solution for 3 minutes
11. Wash in tap water for 3 minutes
12. Tone in gold chloride solution for 10 minutes
13. Rinse in distilled water
14. Reduce in potassium metabisulfite solution for 1 minute
15. Fix in sodium thiosulfate solution for 1 minute.
16. Wash in tap water for 2 minutes.
17. Dehydrate in 95% alcohol, absolute alcohol and clear in xylene 2changes
18. Mount in DPX.

Results

Reticulinfibers – black

Background – grey

11.15.3 Mcmanus for Glycogen (PAS)

Staining and identification of the various types of carbohydrates (polysaccharides &mucopolysaccharides)

Principle

Tissue structures like liver & heart, striated muscles are studied by Periodic acid Shiff stain. Periodic acid reacts with aldehyde group of thecarbohydrates and afterwards reaction with the schiff's reagent produces a red or purple red color.

Reagents

1. 0.5% w/v periodic acid solution
2. Schiff's reagent
   a) Dissolve 1.0 g of basic fuchsin in ≈100 ml of boiling distilled water cool to about 60 degrees and filter.
   b) Add 20 ml of 0.1 N hydrochloric acid cool further and add 1.0g of sodium metabisulphite and mix well.

c) Keep in the dark for 24-48 hours. When the solution become straw colored, add 300 mg of activated charcoal, shake vigorously, filter and store.

4. 1 N Hydrochloric acid
5. 0.1 g of light green in 100 ml of 0.1% (v/v) acetic acid
6. Harris haematoxylin stain

Procedure

1. Deparaffinize and hydrate to distilled water.
2. Oxidize in periodic acid solution for 5 minutes
3. Rinse in distilled water
4. Schiff's regent solution for 15 minutes.
5. Wash in running water for 10 minutes for pink color to develop.
6. Harris haematoxylin for 6 minutes or light green counter stain for a few seconds.
7. Wash in running water.
8. Differentiate in 1% acid alcohol solution 3-10 quick dips.
9. Wash in running water.
10. Dip in ammonia water to blue the sections
11. Wash in running water for 10 minutes
12. Dehydrate in 95% alcohol, absolute alcohol, clear in xylene twochanges each.
13. Mount in DPX.

Results

With hematoxylin counterstain

1. Nuclei – blue
2. Glycogen, mucin, hyaluronic acid, reticulin, colloid droplets, amyloid infiltration, thrombi. – Purple red
3. Fungi – Red
4. Background – pale green (with light green counter staining).

11.15.4 Test for Schiff Reagent Solution

Pour a few drops of schiff reagent solution into 10 ml of formaldehyde (37-40%) in a watch glass. It the solution turns reddish purple rapidly it is good. Ifthe reaction is delayed and the resulting color deep blue purple the solutionin breaking down.

Mayer's Mucicarmine Method

Principle

The rationale for specificity of mucicarmine for mucin is not fully understood by the probable mechanism is that the aluminium salts in the solution form achelate compound with carmine, thus producing a net positive charge on the molecule and consequent binding to the tissue polyanions. The compound has a large molecular size and allows that dye complex topenetrate and bind to acidic substrates of low density like mucins otheracidic substances like nucleic acid are of high density hence exclude themucicarmine.

Materials

1. Weigert's Iron haematoxylin solution
   a) Solution A
      - Hematoxylin crystals 1.0 g
      - Alcohol 95% 100 ml
   b) Solutions B
      - Ferric chloride 29% aqueous 4 ml
      - Hydrochloric acid conc. 1 ml
      - Distilled water 95 ML

Working solution – Add equal parts of solution A and solution B

2. Mucicarmine solution
   - Carmine 1 g
   - Aluminium chloride anhydrous 0.5 g
   - Distilled water 2 ml

Mix stain is small evaporating dish. Heat on electric hot plate for 2 minutes, Liquid becomes black and syrupy. Dilute with 100 ml of 50% alcohol and let it stand for 24 hours filler.

Dilute 1 part of mucicarmine solution with 4 parts of tap water for use.

3. 0.25% Metanil yellow solution
   - Melanil Yellow - 0.25 g
   - Distilled water - 100 ml
   - Glacial acetic acid - 0.25 ml

Procedure

1. Deparaffinize and hydrate to distilled water.
2. Working solution of Weigert's Haematoxylin for 7 minutes.
3. Wash in running water for 10 minutes
4. Diluted mucicarmine solution for 60 minutes
5. Rinse quickly in distilled water.
6. Metanil yellow solution for 1 minute.
7. Rinse quickly in distilled water

8. Dehydrate in 95% alcohol, absolute alcohol and clear in xylene 2 changes each.
9. Mount with DPX.

Results

- Mucin – Deep rose to red.
- Capsule of cryptcoccus - deep rose to red.
- Nuclei – Black
- Other tissue elements – yellow

11.15.5 FITE's Method for Acid Fast Organisms

Principle

The mycobacteria are the bacteria which are relatively resistant to staining because of the lipid capsule which surrounds them, when stained by a strong stain (egcarbolfuchsin) they resist decolourization by acid. In this method a red dye carbolfuchsin is forced into the bacteria and other structures with heat and is then removed from other structures with acid or alcohol; tubercle bacillus because of lipid capsule however resists decolourization.

Reagents:

1. Xylene – peanut oil solution
   - Peanut oil - 1 part
   - Xylene - 2 parts
2. ZiehlNeelsen CarbolFuchsin solution
   - basic fuchsin 1 g
   - Phenol 5 g
   - Distilled water 100 ml
   - Absolute alcohol 10 ml
   - Basic fuchsin is dissolved in alcohol and added to phenol in distilled water. Filter before use
3. 1% (v/v) sulphuric acid solution
4. Methylene blue solution 0.25% methylene blue in 1% glacial acetic acid.

Procedure

1. Deparaffinize through 2 changes of xylene peanut oil solution for 12 minutes each.
2. Drain wipe off excess oil & blot to opacity.
3. Carbolfuchsin solution for 30 minute
4. Wash in tap water for 3 minutes & blot dry.
5. Differentiate slides with sulfuric acid solution till sections are faint pink for 1-2 minutes.

6. Wash in running water for 3 minutes 100
7. Counterstain lightly with working methylene blue solution (few seconds).
8. Rinse off excess methylene blue in tap water.
9. Blot and let it stand for few minutes to air dry thoroughly.
10. Dip slides in xylene before mounting.
11. Mount with DPX
12. Results
13. M. Leprae - red
14. Nocardia filaments - blue grey

11.15.6 Gomori's Method for Iron

Principle

Haemosiderin is a brown granular pigment occurring at the site of previous haemorrhage. It is a product of the breakdown of haemoglobin. It reacts with potassium Ferro cyanide in acid medium and yields a Prussian blue color.

Reagents

1. Solution A - 20% hydrochloric acid solution (stock)
   - Hydrochloric acid (conc.) 20 ml
   - Distilled water 80 ml
   - Distilled water 80 ml
2. Solution B - 10% potassium ferrocyanide solution (stock)
   - Potassium Ferro cyanide 10 g
   - Distilled water 100 ml
3. Acidified potassium Ferro cyanide solution. Prepare fresh by mixingequal part of solution A and solution B and leave for 20 minutes
4. Nuclear fast red stain. Dissolve 5.0 g of aluminium sulphate is hotdistilled water and add 0.1 g of nuclear fast red mix well and filter.
   - Add a crystal of thymol as a preservative.

Procedure

1. Deparaffinize and hydrate to distilled water
2. Put the slides in acidified potassium Ferro cyanide solution for 30minutes
3. Rinse in distilled water
4. Counterstain in nuclear fast red solution for 5 minutes.
5. Rinse in distilled water
6. Dehydrate in 95% alcohol, absolute alcohol, and clear in xylene 2 changes each
7. Mount with DPX

Results

- Iron pigments - bright blue
- Nuclei - Red
- Cytoplasm - Light pink.

11.15.7 Von Kossa's Method for Calcium

Principle

The calcium salts in the form of phosphates, carbonates oroxalates occur as components of oxalates occur as components oflaminated concentration in various organs such as the kidneys, urinary bladder, lymph nodes. Von Kossa's technique in a metal substituted for calcium by metallic salt formation with the anion of the calcium salt.

Materials

1. 5% W/v silver nitrate solution (stored in amber coloured bottle)
2. 5% W/v sodium thosulphate solution
3. Nuclear fast red stain. Dissolve 0.1 g nuclear fast red no 100 ml of 5% solution of aluminium sulphate with the aid of heat. Cool filter add grain of thymol as a preservative.

Procedure

(Use control slides and chemically clean glassware)

1. Deparaffinize and hydrate to distilled water
2. Silver nitrate solution expose to bright sunlight or under the light of 100 watt bulb for 60 minutes
3. Rinse in Distilled water
4. 5% sodium thiosulfate solution for 2 minutes
5. Rinse well in distilled water
6. Counter stain with nuclear fast led solution for 5 minutes
7. Rinse in distilled water
8. Dehydrate in 95% alcohol, absolute alcohol and clear in xylene 2 changes each
9. Mount with DPX

Results

- Calcium salt - Black
- Nuclei - Red
- Cytoplasm - Light pink

11.15.8 Hall's Method for Bilirubin

Principle

Bilirubin is oxidized to biliverdin and stains olive green to emerald green depending on the concentration of bilirubin in solution.

Reagents

1. Fouchet's Reagent
   - Trichloroacetic acid - 25.0 g
   - Distilled water 100ml
2. Mix and add 10 % ferric chloride
3. 10% Ferric chloride solution
   - Ferric chloride - 10 g
   - Distilled water - 100 ml
4. Van Gieson's solution
   a) Saturated aqueous picric acid 10ml
   b) 1% w/v acid fuchsin 1.5ml

Procedure

1. Deparaffinize and hydrate to distilled water
2. Fouchet's reagent for 5 minutes
3. Wash in running water, then in distilled water
4. Van Gieson's solution for 5 minutes
5. Dehydrate in 95% alcohol, absolute alcohol and clear in xylene 2 changes each
6. Mount with DPX

Results

- Biliverdin green
- Collagen red
- Muscle Yellow

11.15.9 Removal of Pigments – Melanin

A. Deparaffinge and Hydrate and Rinse in Distilled Water

- 0.25% potassium permanganate – 30-60 min.
- Wash well in water
- 5% oxalic acid till clear 2-5 min.

Wash in tap water followed by rinse in distilled water

Or

B. Deparaffinize and Hydrate to Distilled Water

- Keep in 10 volumes of hydrogen peroxide – 24 hours
- Wash in distilled water

Or

C. 1% Hydrochloric acid overnight wash well in water followed by wash indistilled water.

11.15.10 Removal of Pigments Formalin

1. Deparaffinize and hydrate to distilled water
2. Let it stand in saturated alcoholic picric acid – 3 hours
3. Wash in running water till the yellow colour in removed.

12
Chapter

Analysis of Growth Parameters

The techniques of growth analysis have been extensively used in recent years for better understanding of the physiological basis of yield variation in crop plants. Growth analysis is a physiological probe on the developmental of the crop in chronological sequence to elucidate and account the causes for differences in yield through the events that have occurred at different stages.

12.1 Dry Matter Production and its Partitioning

The dry matter in the vegetative strictures continues even after the onset of reproductive growth. It mainly attributable to biomass of among the genotyopes.

Materials and Methods

- Hot air oven
- Scator
- Electric weighing balance

Procedure

1. All aboveground plants were weighed to obtain fresh weight. Because of the large bulk of material sampled, a sub sampling procedure was necessary to obtain an estimate of dry weight.
2. The plants samples were partitioned into leaves, stem (including petioles) and reproductive part (pod).
3. Stem were chopped into length of 3 to 5 cm.
4. Dried each sample at 80°C in hot air oven for 72 hours.
5. Completely dried samples were weighed.
6. The dry weight of different plant parts was expressed in g per plant basis.

Calculations

$$\text{Dry matter content} = \frac{(\text{Initial weight} - \text{Final weight/})}{(\text{Initial weight})} \times 100$$

12.2 Leaf Area Index (LAI)

The leaves of the plant are normally its main organs of photosynthesis and the total area of the leaves per unit area of land surface called has proposed by Watson (1947) as the best measure of the capacity of crop for producing dry matter. Katiyar

(1980) reported that there was a continuous increase of LAI in all the varieties for a considerable period after the start of reproductive phase.

Materials

- Graph papers
- Leaflets
- Pencil
- Papers etc.

Procedure

1. The leaf area was worked out by graphical method.
2. Place the leaflet on the graph paper and mark the area covered by the leaflet.
3. These leaflets were then collected and dried and were used for computing leaf area per plant on dry weight basis. (As described above).

Calculations

The LAI per plant was calculated by the following formula

$$LAI = \frac{\text{Leaf area/plant}\ (cm^2)}{\text{Land area/ plant}\ (cm^2)}$$

12.3 Leaf Area Ratio (LAR)

The LAR was calculated using the formula of Radford (1967) and expressed in dm^2g^{-1}.

$$LAR = \frac{\text{Leaf area}\ (dm^2)}{c}$$

12.4 Specific Leaf Area (SLA)

The inverse of the specific leaf weight is the specific leaf area and was calculated as follows

$$SLA = \frac{1}{SLW} dm^2g^{-1}$$

12.5 Leaf Area Duration (LAD)

Leaf area duration is the integral of leaf area index over a growth period (Watson, 1952). LAD for various growth periods was worked out as per the formula of Power *et al.* (1967) and expressed in days.

$$LAD = \frac{Li + (Li + 1)}{2} \times (t_2 - t_1)$$

Whereas;

Li = LAI at i^{th} stage

I at $(I + 1)^{th}$ stage

$T_1 - t_2$ = time interval between and $(I + 1)^{th}$ stage (days)

References

1. Effect of nutrients on physiology, yield and yield components and disease incidence in blackgram [vigna mungo (l.) hepper], akshata s. patil, 2013

12.6 Specific Leaf Weight (SLW)

The specific leaf weight indicates the leaf thickness and was determined by the methods of Radford (1967). It is expressed in g dm-2

$$SLW = \frac{\text{Leaf dry weight (g)}}{\text{Leaf area } (dm^2)}$$

12.7 Biomass Duration (BMD)

The BMD was calculated by using the following formula and expressed in g days

$$BMD = \frac{TDM + TDM\ (i+1)\ (t_2 - t_1)}{2}$$

12.8 Crop Growth Rate (CGR)

Crop growth rate is the rate dry matter production per Unit ground area per unit time (Watson, 1952). It was calculated by using the following formula and expressed as g dm^{-2} day^{-1}.

$$CGR = \frac{W_2 - W_1}{(t_2 - t_1)} \times \frac{1}{A}$$

Whereas;

W_1 = Dry weight of the plant (g) at time t_1

W_2 = Dry weight of the plant (g) at time t_2

A = Land area (dm^2)

12.9 Absolute Growth Rate (AGR)

It expresses the dry weight increase per unit time and was calculated by using the formula,

$$AGR = \frac{W_2 - W_1}{(t_2 - t_1)} \text{g day}^{-1}$$

Whereas;

W_2 and W_1 are the total dry weights per plant at t_2 and t_1, respectively.

12.10 Relative Growth Rate (RGR)

It is the rate of increase in dry weight per unit dry weight already present and is expressed in g day-1. Relatively growth rate at various stages was calculated as suggested by Radford (1967)

$$RGR = \frac{\log_e W_2 - \log_e W_1}{(t_2 - t_1)} \text{ g day}^{-1}$$

Whereas;

W_1 = Dry weight of the plant (g) at time t_1

W_2 = Dry weight of the plant (g) at time t_2

12.11 Net Assimilation Rate (NAR)

Net assimilation rate is the rate of dry weight increase per unit leaf area per unit time. It was calculated by following formula of Radford (1967) and expressed as g dm^{-2} day^{-1}

$$NAR = \frac{W_2 - W_1}{(t_2 - t_1)} \frac{\log_e W_2 - \log_e W_1}{(L_2 - L_1)}$$

Whereas;

L_1 and W_1 = Leaf area (dm^2) and dry weight of the plant (g), respectively at time t_1

L_2 and W_2 = Leaf area (dm^2) and dry weight of the plant (g), respectively at time t_2

References

Causton D R. (1991). Annals of Botany 67, 137 4. Plant growth analysis: the variability of relative growth rate within a sample.

Chakravorty N K K., and P S N Sasstry (1983) Mausam 34 (3): 323-326. Biomass production in wheat in relation to evaporative demand and ambient temperature.

Charles Edwards D A (1982). Academic Press, Sydney. Physiological determination of crop growth.

Hurd RG. (1977). Annals of Botany. 41, 779-87.Vegetative plant growth analysis in controlled environments.

Katiyar R P (1980). Indian J. agric. Sci., 50(9): 684-691. Development changes in leaf area index and other growth parameters in chickpea.

Poorter H. (1989). Physiologia Plantarum. 75, 237-44. Plant growth analysis: towards a synthesis of the classical and the functional approach.

Radford P J (1967). Growth analysis formulae. Their use and abuse. Crop *Sci., 8: 171-175.*

Shakaran A (1966). Laboratory Manual for Agricultural Chemistry, Ed. Asia Publishing House, *Bombay-1, pp. 20-23*

Watson D J. (1947). Ann. Bot., 2 41-76. Comparative physiological studies on the growth of field crops.

www.ingramcontent.com/pod-product-compliance
Ingram Content Group UK Ltd.
Pitfield, Milton Keynes, MK11 3LW, UK
UKHW021947270726
14060UKWH00002B/401